DATE DUE

OC 2 03	JE 1 10		
OC 03			
JA 2			
NO 05			
O			
MO 05			
NO 30 05			
MR 8 06			
AP 25 06			
MY 06			
OC			
OC 30			
DE 19 08			
JA 28 09			
AP 20 09			

DEMCO 38-296

THE
CONCISE
ENCYCLOPEDIA
OF THE
HUMAN BODY

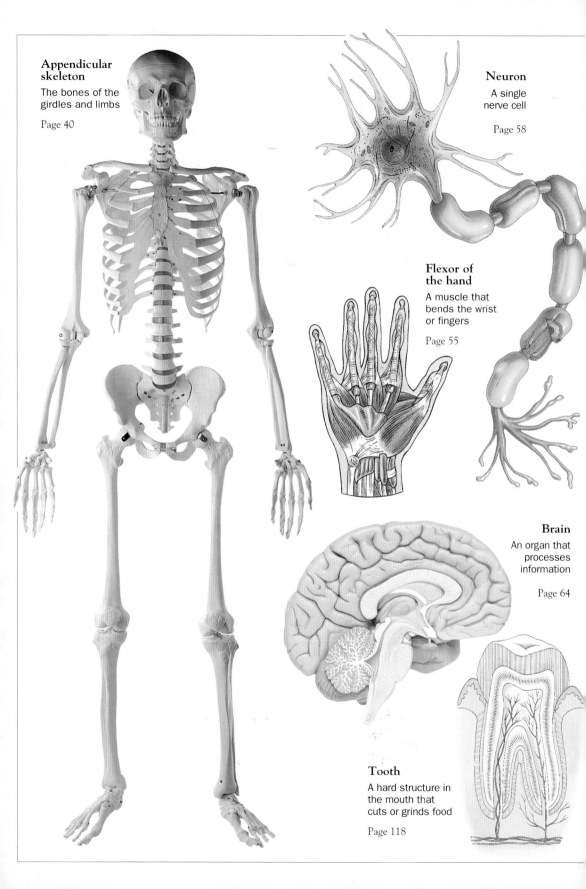

Appendicular skeleton

The bones of the girdles and limbs

Page 40

Neuron

A single nerve cell

Page 58

Flexor of the hand

A muscle that bends the wrist or fingers

Page 55

Brain

An organ that processes information

Page 64

Tooth

A hard structure in the mouth that cuts or grinds food

Page 118

THE
CONCISE
ENCYCLOPEDIA
OF THE
HUMAN BODY

Written by David Burnie

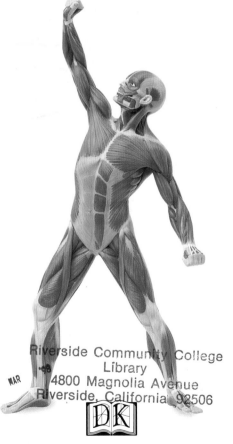

Riverside Community College
Library
MAR 4800 Magnolia Avenue
Riverside, California 92506

DORLING KINDERSLEY
London • New York • Stuttgart

QP37 .B964 1995
Burnie, David.
The concise encyclopedia of
the human body

OK

Project Editors Gillian Cooling & Fiona Robertson

Project Art Editor Mark Regardsoe

Art Editor Ann Cannings

Designer Sarah Cowley

Production Louise Barratt & Ruth Cobb

Senior Editor Stephen Setford

Managing Editor Helen Parker

Managing Art Editor Peter Bailey

Picture Research Giselle Harvey

US Editor Constance V. Mersel

Educational Consultants
Richard Walker, BSc., PhD., PGCE
Kimi Hosoume, B.A., Lawrence Hall of Science,
University of California at Berkeley

Medical Consultant
Dr. Frances Williams MB BChir MRCP DTM & H

Models supplied by Somso Modelle, Coburg, Germany

First American Edition, 1995
2 4 6 8 10 9 7 5 3 1
Published in the United States by
Dorling Kindersley Publishing, Inc.,
95 Madison Avenue, New York, New York 10016
Copyright © 1995 Dorling Kindersley Limited, London
Text copyright © 1995 David Burnie

All rights reserved under International and Pan-American Copyright Conventions. No part
of this publication may be reproduced, stored in a retrieval system, or transmitted in any
form or by any means, electronic, mechanical, photocopying, recording, or otherwise,
without the prior written permission of the copyright owner. Published in Great Britain by
Dorling Kindersley Limited.
Distributed by Houghton Mifflin Company, Boston.

Library of Congress Cataloging-in-Publication Data
Burnie, David.
    Concise encyclopedia of the human body / by David Burnie. – – 1st
American ed.
        p.   cm.
    Includes index.
    Summary: A detailed, well-illustrated guide to the major systems,
functions, and structures of the human body.
    ISBN 0-7894-0204-1
    1. Human physiology – – Encyclopedias, Juvenile. [1. Body, Human –
– Encyclopedias. 2. Human physiology – – Encyclopedias.] I. Title.
QP37.B964   1995
612' .003 – – dc20   95–15134
                                                            CIP
                                                            AC

Reproduced by Colourscan, Singapore
Printed and bound in Italy by New Interlitho

Scientific names

Many parts of the body – bones, for example – have two names. The common name is used in everyday language, while the scientific name is used in human biology and in medicine. In this encyclopedia, scientific names are usually given preference, because they are more precise. If you know only the common name of a body part, look up that name in the index and you will find a cross reference to the scientific name.

Numbers and measurements

Throughout this book, a billion is used to mean one thousand million. Two systems of measurement are used. The first figure is the value expressed in US units. The second figure, which is always in parentheses ( ), is the value expressed in metric, or SI, units. The metric system is generally used for all scientific work.

Contents

HOW TO USE THIS BOOK 8–9

A guide to help you find your way around the Concise Encyclopedia of the Human Body

THE HUMAN BODY 10–11

An introduction to the human body

STUDYING THE BODY 12–19

Terms and techniques used to identify and investigate parts of the body

CHEMISTRY OF THE BODY 20–25

The different kinds of chemicals found in the body

CELLS, TISSUES, & ORGANS 26–31

The body's building blocks, and how they are organized

INTEGUMENTARY SYSTEM 32–33

The system that protects the body from the outside environment

SKELETAL SYSTEM 34–47

The framework that supports and protects the body

MUSCULAR SYSTEM 48–57

How muscles work, and how they are arranged

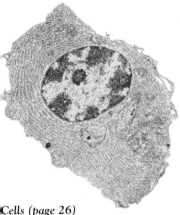

Cells (page 26)

The human body consists of billions of cells, arranged in a highly organized way. Find out about different types of cells, and the way they are arranged, on pages 26–31.

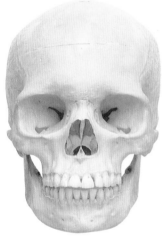

The skull (pages 38–39)

This complex collection of interlocking bones is held together in an unusual way. You can find out more about the skull and the rest of the skeleton on pages 34–47.

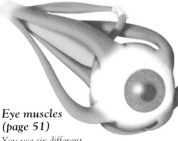

Eye muscles (page 51)

You use six different muscles to move each of your eyes. One of them has a special pulley system – the only one found in the human body. You can find out how eye muscles work on page 51.

NERVOUS SYSTEM 58–67

The body's high-speed communication system

SENSES 68–75

How the body senses changes in the surrounding environment

HOMEOSTASIS 76–77

Built-in controls that keep the body in a stable state

ENDOCRINE SYSTEM 78–81

The system of chemical messengers that influences the way the body works

CIRCULATORY SYSTEM 82–91

The liquid conveyor belt that delivers nutrients and removes waste

DISEASE & IMMUNITY 92–101

What causes disease, and how the body fights back

NUTRITION & METABOLISM 102–109

Essential materials, and how they are used by the body

RESPIRATORY SYSTEM 110–115

The system that allows the exchange of gases between air and blood

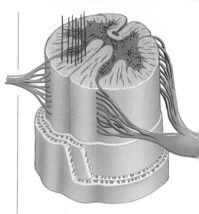

Spinal cord (page 62)
The spinal cord carries millions of electrial signals to and from most parts of your body and allows you to react almost instantly if you touch something hot or sharp. Find out how it works on pages 62–63.

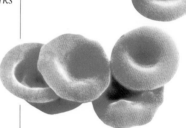

Blood (page 82)
Blood is an extremely complex substance. You can discover exactly what this remarkable fluid contains, and what it does, by looking at pages 82–91.

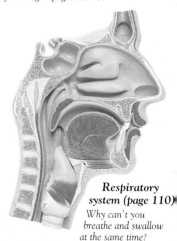

Respiratory system (page 110)
Why can't you breathe and swallow at the same time? Exactly what happens when you sneeze or cough? Find out about breathing and the respiratory system on pages 110–115.

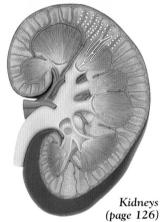

**Kidneys
(page 126)**

Your kidneys are your body's filtering system. Through a complex system of tubes, they allow your body to get rid of cellular waste, while keeping what it needs. You can discover how they work on pages 126–127.

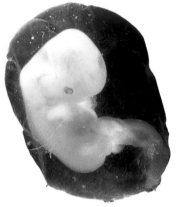

Human embryo (page 137)

It takes about nine months for a single cell to develop into a baby that is ready to be born. You can discover how this cell changes, and how a baby develops, on pages 136–139.

**Francis Crick and
James Watson
(pages 148 & 151)**

These two scientists helped to show how the body's chemical instructions are stored and passed on. Find out more about them, and other doctors and biologists, on pages 148–151.

DIGESTIVE SYSTEM 116–125

The system that breaks down food into a form that can be used

URINARY SYSTEM 126–127

How cellular waste is removed and disposed of

REPRODUCTIVE SYSTEM 128–131

The system that produces new life

INHERITANCE 132–135

How characteristics are passed from parents to their children

GROWTH & DEVELOPMENT 136–143

The pattern of changes that take place during life

TABLE OF INFECTIOUS DISEASES 144

Diseases that are caused by bacteria, viruses, and other organisms

TABLE OF NONINFECTIOUS DISEASES & DISORDERS 146

Diseases that are triggered by genes, or by factors in the world around us

PIONEERS OF HUMAN BIOLOGY & MEDICINE 148–151

More than 100 of the world's greatest doctors and biologists

INDEX 152–159

More than 2,000 key words, terms, and concepts used in human biology

ACKNOWLEDGMENTS 160

How to use this book

This encyclopedia explains and illustrates the most important words and concepts in human biology. It is a thematic encyclopedia, with words arranged into subject areas, such as "Nerves" and "Muscles." This enables you to find out about a whole subject, as well as about individual words. To look up a word, turn to the index at the back of the book. To look up a subject, you can either look in the index, or turn to the Contents on pages 5–7. This lists the different sections and subjects covered in this book.

Main illustration
A large photograph or piece of art usually illustrates several related entries. It helps to explain the entries or to show how they are linked. This picture of a neuron helps to explain its structure.

Cross reference
A small gray square (■) after a word shows that the word is a main entry or a subentry elsewhere in the encyclopedia. The "See also" box gives the page number of the entry.

Subentries
A subentry is printed in bold type. It gives the meaning of a word that is related to the main entry. This subentry explains the meaning of the word "dendrite."

Using the index
The index lists all the entries in alphabetical order and gives their page numbers. If you look up nerve in the index, for example, you will find that the entry is on page 58. The word you want may be a main entry, or it may be a subentry, which is found in bold type within an entry. It may also be an entry in a table.

Subject heading and introduction
A main heading introduces the subject. All the entries in this subject are concerned with nerves. Each subject begins with an introduction that gives you a brief overview of what follows.

---

58 • Nervous System

Nerves

Nerves enable the different parts of your body to work as a single unit. They allow you to take in information about the world around you and to react quickly to changes. They also control many of your body's internal processes.

Nerve
A bundle of cells that carry signals

Nerves are the body's "wiring." Each one is covered by a tough, fibrous sheath and contains hundreds or thousands of cells called neurons. Nerves run out from the brain ■ and spinal cord ■. They reach all parts of the body, including the skin, muscles, and sense organs ■, and even inside teeth and bones.

Neuron
A single nerve cell

Neurons are also known as nerve cells. They are adapted for carrying electrical signals, called nerve impulses. All neurons have a cell body, containing a nucleus ■, short filaments called dendrites, which carry electrical signals towards the cell body, and a long filament called an axon, which usually carries the signals away. Unlike most cells, neurons cannot divide once formed. As the body grows, neurons that die are not replaced, and the number gradually decreases.

Axon
A long filament in a neuron

An axon, or nerve fiber, is the part of a nerve cell that actually carries an electrical signal, or nerve impulse. Axons are finer than a hair, and although many are less than 0.04 inch (1 mm) long, some are over 3 ft (1 m) long. They are often surrounded by a sheath of a fatty substance called myelin. Myelin works like the plastic around an electrical wire, and helps the fast flow of nerve impulses.

Synaptic knob

Axon

Myelin sheath formed by glial cells

Node of Ranvier

Dendrite

Cell body

Nucleus

Structure of a neuron
A typical neuron has a cell body, which contains the cell's nucleus, and a long axon. The axon is insulated by a myelin sheath. Between the glial cells that form the myelin sheaths are spaces called nodes of Ranvier.

Glial cell
A cell that supports, protects, nourishes nerve cells

Glial cells do not carry nerve impulses. Instead, they support nerves by providing them with nutrients, or by attacking invading bacteria ■. Special cells called Schwann cells wrap themselves around the axons of some neurons and are rich in insulating substance myelin.

Stimulus
A physical or chemical change that affects a nerve

A stimulus is anything that changes the electrical state of a neuron. If the stimulus is above a certain level, called a neuron's threshold level, it fires off a nerve impulse, producing a response. Stimuli are very varied. External stimuli come from outside the body and include changes in temperature, pressure, or light intensity. Internal stimuli come from inside the body. They include changes in osmotic pressure and hormone levels.

Captions and annotations
A headed caption explains what you can see in a picture. The caption to this picture describes the structure of a typical neuron. Details in a picture, such as the neuron's cell body, are pointed out by annotations.

Definitions
The definition is a short, precise description. This definition tells you what a neuromuscular junction is.

Explanations
The explanation provides more information about the entry. It can help you to understand the definition and the illustration (if any). It also shows how this entry is linked to others in the same subject. This explanation describes the role of a neuromuscular junction in nerves.

Running heads
For quick reference, the running head shows you which section you are in. This section is the Nervous System.

Biographies
On page 59, you will find a biography of Camillo Golgi, who played a key role in finding out how nerves work. This encyclopedia contains many biographies of famous scientists, linked to the subjects that they investigated. There is also a fuller alphabetical list of famous biologists and doctors on pages 148–151.

Photographs
Detailed photographs reveal structures inside and outside the body.

Three-dimensional models
In some places, special models are used to show living structures. This model shows the different parts of the brain.

in entry ...dings
...entry is ...ve impulse.

...ve impulse
...al that passes down a neuron

...ing impulse is rather like a ...ed battery. It uses an active ...ort ■ system called the ...n-potassium pump to ...sodium ions ■ out through ...sma membrane ■, and to ...potassium ions in. This ...nt pumping creates a tiny ...cal charge. If the cell ...es a stimulus, sodium ions ...back across the membrane. ...everses the charge in that ...nd an electrical ...bance, or **action potential**, ...s along the axon at up to ...(100 m) per second. ...n the disturbance reaches a ...se, it can trigger an impulse ...adjacent nerve.

...apse
...tion between two neurons

...apse is a junction that ...s one neuron to trigger an ...se in another. It consists of ...ll swelling at the end of an ...**synaptic knob**, ... When a nerve imp... ...es the synaptic knob, it ...rs the release of a substance ...a **neurotransmitter**. This ...es the gap between the two ...and in less than a ...econd, triggers the second ...into action. Some neurons ...hundreds or thousands of ...ses. A synapse always passes ...in the same direction; it ...ot work in reverse.

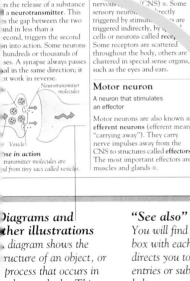

Neurotransmitter molecules
Vesicle

Synapse in action
...transmitter molecules are ...d from tiny sacs called vesicles.

Neuromuscular junction
A special synapse between neurons and muscle fibres

A neuromuscular junction is a synapse between a motor neuron and a muscle fiber ■. When an impulse arrives at the synapse, it makes the muscle ■ contract.

A neuromuscular junction
This false color electron micrograph shows a motor neuron (pink) attached to skeletal muscle fibers.

Sensory neuron
A neuron that carries signals to the central nervous system

Sensory neurons are also called **afferent neurons** (afferent means "carrying toward"). They carry ...rve impulses from different ...body to the central nervous ...(CNS) ■. Some sensory neuron... ...ectly triggered by stimuli... ...triggered indirectly, by s... cells or neurons called **rece**... Some receptors are scattered throughout the body, others are clustered in special sense organs, such as the eyes and ears.

Motor neuron
A neuron that stimulates an effector

Motor neurons are also known as **efferent neurons** (efferent means "carrying away"). They carry nerve impulses away from the CNS to structures called **effectors**. The most important effectors are muscles and glands ■.

Camillo Golgi
Italian histologist (1843–1926)

Camillo Golgi made an important breakthrough that allowed nerve cells to be seen clearly for the first time. He devised a technique that stains nerve cells black, but leaves the surrounding cells almost unaffected. Using this stain, he identified different types of nerve cell, and found that synapses contain small gaps between one cell and the next. He also discovered the Golgi body ■, an organelle ■ that is present in most cells.

Association neuron
A neuron that passes signals from one neuron to another

Association neurons enable the nervous system to pass on signals, and to sort and compare them. These neurons are found mainly in the central nervous system.

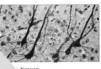

...power
...micrograph shows association neu... the brain.

See also
Activ... ransport 28 • Bacteria 92 ... Brain 64 • Cell division 30 ...ntral nervous system 60 • Gland 78 ...G... body 26 • Ion 21 • Membrane 2... Muscle... Muscle fiber 48 • ... Organelle 2... ...ssure 29 ... Sense organ 68 • Spinal cord 62

...n brain is the control center for the ...ontains billions of nerve cells. When ...rking hard, these cells use about 20 ...he body's oxygen. The brain is ...nto several "departments," each ...ifferent part of the body.

Brain

Forebrain
A part of the brain that deals with homeostasis, emotions, and conscious actions

Most of the forebrain consists of the cerebrum, which is where conscious thought takes place. Hidden beneath the cerebrum are two much smaller structures – the hypothalamus and thalamus.

Thalamus — Cerebrum — Forebrain — Frontal lobe — Optic chiasma — Hypothalamus — Pituitary gland — Midbrain — Pons — Brain stem — 4th ventricle — Cerebellum — Medulla oblongata — Spinal cord

Inside the brain
The model above shows a cross section through the brain, as if the head is facing to the right.

...n brain that merges ...cord

...n is the body's equivalent of an autopilot. It controls the basic processes that are essential for life, such as breathing, rate of heartbeat, and blood pressure. The lowest part of the brain stem is the **medulla oblongata**, or **medulla**. The medulla maintains these vital functions, and is the point where many nerve fibers cross over. Higher up is a swelling called the **pons**, which means "bridge." The pons relays information between the brain and the spinal cord. The highest part of the brain stem is the **midbrain**, which is involved in many reflexes ■.

Cerebellum
The part of the brain that coordinates subconscious movements

The cerebellum ensures that your body is balanced and moves in a coordinated way. It consists of millions of neurons ■ packed into two folded halves, or **hemispheres**, and makes up about 10 percent of the brain's weight. The cerebellum constantly receives "updates" about the body's position and movements. By sending instructions to muscles, it adjusts the body's posture and keeps it moving smoothly.

Grey matter — Cerebrum — White matter — Thalamus — Pons — Cerebellum — Medulla oblongata

Frontal section through brain

Diagrams and other illustrations
...diagram shows the ...ructure of an object, or ...process that occurs in ...e human body. This ...iagram shows the ...ructure of a synapse, ...hich is a junction ...etween two neurons.

"See also" boxes
You will find a "See also" box with each subject. This directs you to other main entries or subentries that can help you to understand the subject better. This "See also" box points to some of the structures that are important in the nervous system.

The human body

Today, there are about six billion people on the Earth. Despite this huge total, everybody – unless they have an identical twin – is unique. We differ in size, height, and shape. The color of our skin, hair, and eyes varies between different individuals and different races. Our fingerprints are unique, as are our voices and even the way we walk. Variations like this are found in all living things, but we notice them more easily in ourselves than in the animals or plants around us. Despite these variations, the human body is always built to a set pattern, and works in the same way. Everyone's body contains the same kinds of cells, the same set of bones, the same groups of muscles, and the same control systems. Without them, the body would not work.

Kill or cure
Early medical treatment often involved mistakes in the two key areas of medicine – diagnosis and therapy (see p.93). This engraving, from the early 15th century, shows a person having a hole cut in their skull. This drastic form of therapy was meant to let evil spirits escape from the body.

Animal allies
Leeches live on blood. When they feed, they produce an anticoagulant (see p.84), which helps to keep the blood flowing. Until the 19th century, leeches were used in Western medicine. Today, the use of artificial anticoagulants is more common.

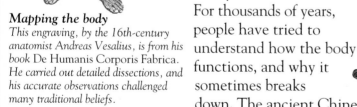

Mapping the body
This engraving, by the 16th-century anatomist Andreas Vesalius, is from his book De Humanis Corporis Fabrica. He carried out detailed dissections, and his accurate observations challenged many traditional beliefs.

Early ideas

For thousands of years, people have tried to understand how the body functions, and why it sometimes breaks down. The ancient Chinese studied the body in great detail, and in the time of the Pharaohs, in about 3,000 BC, the Egyptians had special schools for training priests, who also acted as doctors. But early ideas about the body were shaped as much by myths and superstition as by careful study. People had little real knowledge about the body's different parts, and curing diseases was often a matter of luck rather than skill.

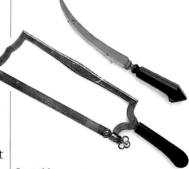

Invisible enemies
All surgical instruments are now sterilized to keep them free of bacteria. Before this technique was introduced in the 19th century, surgery often caused as much disease as it cured.

Hidden world
During the 19th century, improved microscopes (see p.16) allowed scientists to investigate the fine structure of tissues (see p.18). New chemical stains (see Camillo Golgi p.59) that clearly revealed individual cells were also devised during this century.

Anatomical drawings

One of the first people to look at the body in a scientific way was the Greek physician Hippocrates (c.460–377 BC). However, it was not until two thousand years later that the first truly accurate study of the body appeared. Called *De Humanis Corporis Fabrica*, or "On the Fabric of the Human Body," it was written by Andreas Vesalius (1514–64), a Belgian who became professor of anatomy in Padua, Italy, when he was 24. His book showed many detailed drawings of parts of the body, and revealed for the first time exactly how those parts fit together.

Hidden worlds revealed

In the 15th century there were no microscopes, so Vesalius could only describe what could be seen with the naked eye. But with the invention of the light microscope in the late 17th century, and the electron microscope in the 20th century, scientists have been able to enter the hidden world of cells – the tiny units that make up all living things, including ourselves. Today, we know not only what cells look like, but how they work, and how they pass on their chemical instructions when a new human body comes into being. As more and more is learned about the body, more and more terms are used to name its parts and the processes that occur inside it. This book is designed to help you make sense of those words, so that you can understand how your own body is put together, and how it functions.

All-seeing eye
In the 20th century, the electron microscope has enabled scientists to look at individual cells in great detail. Electron micrographs help to show how cells interact with others, and how they change during disease, for example, if they become cancerous (see p.31).

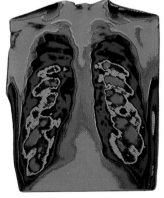

The body at work
New imaging techniques – such as angiography, shown above – reveal internal body structures. This false color angiograph shows the pulmonary circulation (see p.90) within the chest.

The virtual body
As well as examining real bodies, scientists and doctors can now study "bodies" created by computer. This computer-generated picture shows some of the body systems (see p.18) that work together to sustain life.

Studying the body

The human body is the most intensively studied subject on the Earth. Scientists specializing in different areas investigate how the body is made up and how its different parts work.

Biology

The study of life

Biology is the study of all living things. **Human biology** is the study of human beings. Humans are one form of life, or **species**, out of the many millions that live on the Earth.

Anatomy

The study of the body's structure

To understand how the body works, scientists and doctors need to know the shape of its parts and how they fit together. At one time, internal anatomy could only be studied by **dissection** – that is, cutting open dead bodies. Today, an **anatomist** can study the structure of living people with modern imaging ■ techniques.

Physiology

The study of the way the body works

Like everything else on the Earth, the human body obeys the laws of physics. A **physiologist** uses physics to investigate processes such as gas exchange ■, the movement of nerve impulses ■, and thermoregulation ■.

Bacterial samples
This pathologist is looking at cultures of bacteria from the body under a microscope.

Pathology

The study of disease

A **pathologist** studies diseases ■ and their effects on the body. An **oncologist** investigates the causes and development of cancer ■, while an **epidemiologist** probes the causes and spread of infections ■ and diseases.

Histology

The study of tissues

A **histologist** studies the detailed structure of tissues ■. Most of this work involves using microscopes ■, and special stains that make parts of cells ■ more easily visible. **Cytology** is the study of individual cells.

Biochemistry

The study of the chemistry of the body

A **biochemist** studies the chemicals in the body, and finds out how they react with each other. Some of this work can be done in a laboratory by imitating conditions inside the body; some is carried out on living people.

Testing the body
By measuring chemical levels or energy output during exercise, scientists find out more about how the body works.

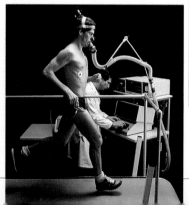

See also
Body system 18 • Cancer 31
Cell 26 • Disease 92
Gas exchange 115 • Infection 92
Magnetic resonance imaging 17
Microscope 16 • Nerve impulse 59
Organ 18 • Thermoregulation 76
Tissue 18

SPECIALIZED STUDY

The study of a particular body system ■ or organ ■ often has its own specific medical name.

Cardiology
The study of the heart
(see pp. 86–87)

Dermatology
The study of the skin
(see pp. 32–33)

Endocrinology
The study of the endocrine system (see pp.78–81)

Gastroenterology
The study of the digestive system (see pp. 116–125)

Gynecology
The study of the female reproductive system
(see pp. 128–131)

Hematology
The study of blood
(see pp. 82–85)

Immunology
The study of the immune system (see pp. 92–101)

Neurology
The study of the nervous system (see pp. 58–67)

Ophthalmology
The study of the eyes
(see pp. 68–71)

Osteology
The study of bones and the skeleton (see pp. 34–47)

Word roots

Many of the terms that appear in this book are made up of parts of words, or "roots," that come from either Latin or Greek. The table below lists some of the most common word roots, and explains what they mean.

Word root

A Latin or Greek word that forms part of a biological term

A single term may have several word roots. For example, the term hepatocyte has two roots – "hepato," meaning liver, and "cyt," meaning cell. Therefore, a hepatocyte is a cell in the liver.

LATIN AND GREEK WORD ROOTS

Word root	Meaning	Example	Page reference
ab	away from	abductor	46
ad	near, towards	adductor	46
anti	against or opposed to	antibody	98
bio	life	antibiotic	95
brachi	arm	brachial artery	91
cardi	heart	cardiovascular	90
cyt	cell	lymphocyte	83
derm	skin	epidermis	32
di	two, twice	diploid cell	133
epi	upon, all over	epithelium	19
ferent	carrying	efferent neuron	59
gastro	stomach	gastric juice	121
gen	a cause of something	antigen	98
genesis	formation	oogenesis	129
hemo	blood	hemoglobin	82
hepato	liver	hepatocyte	122
homeo	similar	homeostasis	76
homo	same, identical	homologous	133
leuco	white	leucocyte	83
macro	large	macromolecule	20
meso	in the middle	mesoderm	137
micro	small	microorganism	92
mono	one, single	monosaccharide	22
myo	muscle	myosin	48
nephro	kidney	nephron	126
neuro	nerve	neuron	58
osteo	bone	osteocyte	34
patho	disease	pathogen	92
pect	chest	pectoral girdle	40
pelv	basin	pelvic girdle	41
peri	near, around	periosteum	35
phago	to eat	phagocytosis	29
plasm	living matter	cytoplasm	26
poly	many	polysaccharide	22
pulmo	lung	pulmonary artery	90
ren	kidney	adrenal gland	81
scler	hard	sclera	68
tri	three, thrice	triglyceride	23

Background picture: lymphocyte

Regions of the body

Scientists and medical practitioners need precise terms to refer to regions of the body, and to show exactly where things are. Here, you can find out about some of these important signposts.

Head

The part of the body that houses the brain

The head houses the brain and many of the body's sense organs ■. It is protected by the skull ■, a framework of interlocking bones that surround the brain and support the **face**. The head is held upright by bones and muscles in the **neck**. Two special vertebrae ■ at the top of the backbone allow the head to tilt and swivel. The word **cephalic** describes something in the head. The word **cervical** describes something in the neck, or the cervix ■.

Trunk

The central part of the body that houses the heart, lungs, and digestive system

The trunk is roughly divided into two halves. The **thorax**, or **chest**, forms the upper part of the trunk and runs from the base of the neck to the diaphragm ■. It contains the heart and lungs, which are protected by the rib cage ■. The thorax can change shape. This movement allows air to be drawn into the lungs. The word **thoracic** describes anything found in the thorax. The **abdomen** forms the lower part of the trunk. Most of it lies below the rib cage, and its organs are protected by layers of muscle instead of by bones. The abdomen contains most of the organs of the digestive system ■, together with other organs such as the kidneys ■ and bladder ■. The word **abdominal** describes anything found in the abdomen.

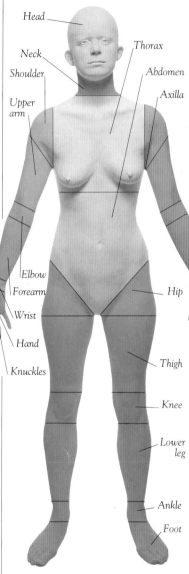

Head
Neck
Shoulder
Upper arm
Thorax
Abdomen
Axilla
Elbow
Forearm
Wrist
Hand
Knuckles
Hip
Thigh
Knee
Lower leg
Ankle
Foot

Key
☐ Head   ■ Neck   ■ Trunk
☐ Upper extremity ☐ Lower extremity

Major parts of the body
The main regions of the body are shown in the diagram above.

Upper extremity

An arm

Extremity is the anatomical term for a limb. Each arm contains 30 bones, and is divided into three regions – the **upper arm**, the **lower arm**, or **forearm**, and the hand ■. The bones meet to form many different joints ■, including the shoulder ■, the elbow ■, the wrist ■, and the knuckles ■. Together, these give the arm amazing flexibility. The armpit, or **axilla**, is found below the point where the arm meets the trunk.

Lower extremity

A leg

Each leg contains 30 bones, and is divided into three regions – the **upper leg**, or **thigh**, the **lower leg**, and the foot ■. The joints in the leg include the hip ■, the knee ■, and the ankle ■.

Cavity

A closed chamber inside the body

Many of the body's organs are found in closed chambers called cavities. The **cranial cavity** contains the brain, while the **thoracic cavity** contains the heart and lungs. The **abdominal cavity** contains most of the digestive system. The organs within a cavity often "float" in a thin jacket of fluid. This fluid acts as a shock absorber, and enables different organs to slide over each other easily.

Cranial cavity
Thoracic cavity
Abdominal cavity

Body cavities
There are three major cavities within the body.

xis

n imaginary line that runs down
e center of the body

he body's axis, or **midline**, splits
in two. Humans are not exactly
mmetrical, so the two halves
e slightly different. This is most
oticeable inside the body. Many
gans, such as the liver ▪, lie to
ne side, and some paired organs,
ch as the lungs, are slightly
nequal in size. Humans are also
ymmetrical from the outside.
lany people have one
e or ear higher
an the other,
nd one foot
ightly bigger
an the other.

Medial

pper, middle,
nd lower

arious words
e used to
scribe
hether a part
the body is above,
low, nearer, or
rther from the
idline than another
rt of the body.

Inferior

uperior

oward the head, or toward the
pper end of the body

n anatomy, the word superior
oes not indicate that one thing
better than another. It means
at something lies higher up
an another part of the body.
or example, the adrenal glands ▪
re superior to the kidneys. The
ord superior can also form part
a name. For example, the
perior vena cava ▪ is a vein
at drains blood from the upper
ody, and the superior rectus
uscle ▪ is one of the uppermost
uscles that swivel the eye.

Inferior

Toward the feet, or toward the
lower end of the body

When used as an anatomical
term, the word inferior indicates
that one part of the body is lower
than another part. For example,
the stomach is inferior to the
diaphragm, because the
stomach is below. It is
still inferior even
if you stand on
your head,
because all
directional
terms are taken from
a standing position. Like the
word superior, the word inferior
can also form part of a name. For
example, the inferior vena cava
is a large vein that drains blood
from the lower part of the body.

Medial

At or nearer the midline

Something that is medial lies near
the body's midline. A part that is
lateral lies to the side of it.

Superficial

At or near the surface

A superficial structure is one
close to the outside of the body,
or to the outside of an organ. For
example, the epidermis ▪ is a
superficial layer of the skin, and
the sclera ▪ is a superficial layer
of the eye. A **deep** part is one
that is located away from the
surface of the body. For example,
the ribs are deep in relation to
the skin of the chest.

Peripheral

At or near the farthest regions
of the body

Something that is peripheral lies
away from the center of the body.
The peripheral nervous system,
for example, contains nerves that
reach to the fingers and toes.

Anterior

Toward the front of the body

An anterior part of the body is
one that lies in front of something
else. For example, the heart is
anterior to the backbone. The
front surface of the body is called
the anterior,
or **ventral**,
surface.

*Front,
back, far,
and near*
*Directional
terms are used
to locate body
parts in relation
to one another.*

Posterior

Toward the back of the body

A posterior part of the body is one
that lies behind something else.
For example, a spinous process ▪ is
posterior to the rest of a vertebra
because it projects behind it. The
back surface of the body is known
as its posterior, or **dorsal**, surface.

Proximal

At or near the point of attachment

Many parts of the body, such as
the arms, fingers, and nails are
attached at one end. A proximal
part is one near the point of
attachment, while a **distal** part is
one further away from it.

See also

Inside the body

With the help of modern technology, it is now possible to see deep inside a living body. Some imaging techniques are good at showing bones or teeth, while others reveal internal organs.

Microscopy

A viewing technique that involves magnifying very small objects

The body's cells ■ are much too small to be seen by the naked eye. To make them visible, they have to be magnified hundreds of times using an instrument called a **microscope**. Pictures produced using microscopes are called **micrographs**. In a **light microscope**, light rays pass through a thin slice of tissue ■, and are focused by lenses to produce a magnified image, or a **light micrograph**. An **electron microscope** uses a beam of electrons ■ to produce pictures called **electron micrographs**. Electron microscopes are more powerful than light microscopes, but only produce black-and-white images. In this book, many of the electron micrographs have been artificially colored by computer.

Electron micrograph
This false color electron micrograph shows a macrophage cell.

Endoscopy

A technique used for seeing inside the body's cavities and organs

In endoscopy, a flexible viewing instrument called an **endoscope** is introduced into a space inside the body. The endoscope lights up its surroundings, and forms an image of them in a hand-held viewer. Endoscopy is often used to examine the stomach and other parts of the alimentary canal ■. A similar technique called **arthroscopy** is used for examining joints ■.

X-ray
X-ray pictures are useful for investigating structures inside the body. This X-ray is of the chest.

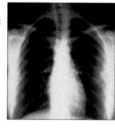

Radiography

A method of examining the body that uses intense radiation

Radiography uses an intense form of radiation called **X-rays**. X-rays can pass through the soft parts of the body, but are absorbed by dense materials such as bone ■ or dentine ■. When they emerge from the body, the X-rays hit a photographic film. The film turns gray or black where X-rays have passed through the body, but stays white where they have been absorbed. This type of X-ray picture is called a **radiograph**. X-rays can damage living cells, so they have to be used with care.

Angiography

The examination of blood vessels using a dye that absorbs X-rays

In an **angiograph**, blood vessels are injected with a harmless dye that absorbs X-rays. As the dye travels through part of the body, X-ray pictures called **angiograms** are taken to reveal the outline of arteries or veins.

Angiogram
On this false color angiogram of the heart, the main arteries appear as yellow lines.

Computed tomography

A method of examining the body that uses X-rays taken from many angles

In a **computed tomograph** (CT scan, a machine slowly rotates around a part of the body, and fires bursts of X-rays through it. A computer analyzes the way the X-rays are absorbed, and uses this to build up a very precise picture. CT scans are much more detailed than ordinary X-ray pictures, and they reveal all the tissues in a cross section of the body.

Barium meal

A liquid used to investigate the alimentary canal

Barium is a dense element ■ that absorbs X-rays. In a barium meal or **barium swallow**, a compound of barium is mixed to make a white liquid. The patient swallows the liquid, and X-ray pictures are taken as the liquid passes through the digestive system. The X-rays are absorbed by the barium, producing a clear outline of the alimentary canal.

Magnetic resonance imaging

A method of examining the body that uses magnetism and radiowaves

In a magnetic resonance image, or **MRI**, a person's entire body is slid into a machine that produces a powerful magnetic field. The field lines up the minute particles inside the body's hydrogen atoms ■, and bursts of radiowaves are then used to jolt them out of alignment. Each time this happens, the atoms produce signals that are detected and processed by a computer. The computer can store signals from the whole body, and can produce a picture of any part of it at the touch of a button.

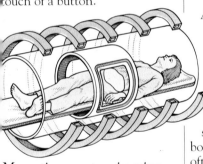

Magnetic resonance imaging
MRI is a harmless and painless procedure. In this diagram, it is being used to provide a detailed cross section of the abdomen.

Ultrasound scanning

A method of examining part of the body using sound

An ultrasound scanner works by beaming pulses of high-frequency sound at the body. It then analyzes the echoes that are reflected back, and uses them to build up a picture. Ultrasound scanning is very safe because it does not use radiation. It is used mainly to examine an unborn fetus ■ in the womb for abnormalities, and is usually performed at about 16–18 weeks of pregnancy ■.

Thermography

A technique for viewing heat radiated by the body

A **thermograph** is a color-coded image that shows the surface temperature of different parts of the body. Thermography can be used to detect areas that have an abnormal temperature, which can sometimes indicate circulatory system ■ problems, or the presence of cancerous ■ cells.

Positron emission tomography

A method of examining the body that uses radioactive chemicals

A positron emission tomograph (**PET**) scanner tracks harmless radioactive chemicals that are injected into the body. Unlike most other kinds of imaging, PET scanning produces pictures that show parts of the body at work. It is often used to examine the brain.

Biopsy

The removal of a small piece of tissue for examination

When doctors suspect that part of someone's body is diseased, they may rely on a biopsy to confirm their diagnosis ■. During a biopsy, a small piece of tissue is cut away and removed. The cells can then be examined under a microscope to see if they show any abnormalities. An **autopsy** is a detailed examination of a person who has died. It is carried out to establish the cause of death.

Ophthalmoscope

An instrument used to examine the inside of the eyes

An ophthalmoscope shines a bright beam of light into the eye, so that the retina ■ can be seen. Disorders of the retina are sometimes a sign of disease elsewhere in the body.

Stethoscope

An instrument used to hear sounds made by the heart and lungs

A stethoscope gathers sounds from a person's chest, and makes them easy to hear. Doctors use stethoscopes to listen for unusual sounds during breathing or while the heart is beating. A standard stethoscope has a flexible plastic tube with two earpieces. At the base of the stethoscope is a sound-collecting device. One side of this device has a thin, plastic diaphragm; the other side is a concave bell with a hole in its center. A stethoscope is also used to help measure blood pressure ■.

Earpieces

Tubing

Bell

Stethoscope
The bell detects low-pitched sounds; the diaphragm detects high-pitched sounds.

Hierarchy of the body

The human body is organized into a sequence, or hierarchy, of different levels. Body systems are made up of a number of different organs, which in turn are made up of tissues and cells.

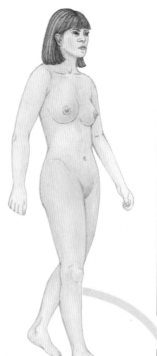

1 Human body
The human body is organized into several different levels.

BODY SYSTEMS

System	Page no.
Integumentary system	32–33
Skeletal system	34–47
Muscular system	48–57
Nervous system	58–67
Endocrine system	78–81
Circulatory system	82–91
Lymphatic system	96–97
Immune system	98–101
Respiratory system	110–115
Digestive system	116–125
Urinary system	126–127
Reproductive system	128–131

Background picture: human body model

Body system

A collection of organs that work together to perform specific tasks

The body is organized on several levels. At the highest level, there are more than 12 body systems. Each system carries out a process that is essential for life, such as digestion ■, excretion ■, or sexual reproduction ■. These systems are made up of organs. Most organs are involved in just one body system, but a few, such as the pancreas ■, play a part in two. The organs are made of different tissues, which are themselves made of a number of different types of cells.

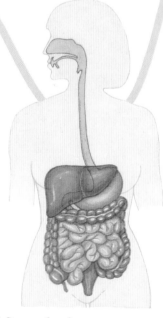

2 System level
The digestive system contains a range of organs that work together.

Organ

A structure made of two or more tissues that carries out a particular range of tasks

The eyes, lungs, kidneys, skin, and stomach are all examples of organs. Each one contains several kinds of tissue, and carries out specific functions that are essential for the body's survival. The cells ■ in organs are specialized for a particular kind of work.

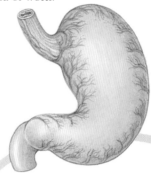

3 Organ level
The stomach is one of the organs found in the digestive system.

Tissue

A group of cells that work together

A tissue contains different kinds of cells that work together to carry out specific tasks. There are many kinds of tissue, but they fall into four categories – epithelial tissue, connective tissue, muscular tissue, and nervous tissue. If some tissue is damaged, for example if you cut your skin, its cells slowly divide and repair the damaged area. This is called **regeneration,** and it results in **tissue repair**. The new cells are often different from the ones that they replace, and may form **scar tissue**, which creates a permanent mark in the area that was damaged. The ability to regenerate varies from one part of the body to another. Connective tissue is good at regenerating, but nervous tissue is not.

Epithelial tissue

A tissue that forms a lining in or on the body

An **epithelium** is a sheet of cells packed closely together. It may be just one cell thick, or it may contain many layers of cells. Epithelial tissue forms the skin 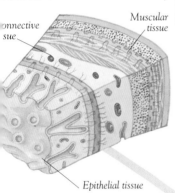, the lining of the alimentary canal, and the lining of the respiratory tract. It produces male and female sex cells, and also forms the inner surfaces of glands. Epithelial tissue protects the body's surfaces. It prevents microorganisms from entering the body; its cells keep dividing so the surface does not wear away.

Connective tissue

Muscular tissue

Epithelial tissue

Tissue level
The stomach contains layers of three kinds of tissue within its walls.

Connective tissue

A tissue that supports the body and holds it together

Connective tissue forms the framework of the body, and shields it from damage. It includes bone and cartilage, which form the skeleton, and several other kinds of tissue that protect and support the body's organs. Many cells in connective tissue produce fibers of collagen and elastin, which give tissue flexibility and strength. **Adipose tissue** is a kind of connective tissue that contains large stores of fat. Blood is a liquid connective tissue that circulates throughout the body.

Muscular tissue

A tissue that produces movement

The cells in muscular tissue contract when triggered by nerves or hormones. When they shorten, they either move part of the body, or change the shape of hollow organs. There are three different kinds of muscular tissue, and all have cells in the form of long fibers. Skeletal muscle is attached to bone, while smooth muscle is found in soft parts of the body. Cardiac muscle is found only in the heart.

Nervous tissue

A tissue that carries electrical signals

Nervous tissue makes up the body's most important communication system. It consists of neurons, which carry nerve impulses, and glial cells, which support neurons and protect them. Neurons are some of the longest cells in the body. Unlike most other cells, they do not divide once they have been formed.

Mucous membrane

A membrane lining part of the body that opens to the outside

Mucous membranes are sheets of cells that line the alimentary canal and the respiratory system. The outermost cells produce a thick fluid called **mucus**. This lubricates the surface of the membrane, and can also trap foreign particles such as airborne dust. All mucous membranes consist of a layer of epithelial tissue backed by a layer of connective tissue. In most mucous membranes, the connective tissue is itself backed by a layer of muscular tissue.

Serous membrane

A membrane lining part of the body that does not open to the outside

Serous membranes line body cavities and surround the organs inside them. They produce **serous fluid**, which lubricates the organs so they can slide past each other easily. Serous membranes are made of epithelial tissue backed by connective tissue. They include the pleural membranes, which surround the lungs, and the peritoneum, which lines the abdominal cavity.

Mucus-producing cell

Acid-producing cell

Hormone-producing cell

Enzyme-producing cell

5 Cellular level
There are many different kinds of cell in the body. Shown above are four kinds of cell found within the stomach lining.

Chemistry of the body

Like all forms of life, the human body is made up of different chemicals. By bringing these chemicals together and causing them to react, the body produces substances it needs to build itself up, and the energy it needs to stay alive.

Atom

A tiny particle of matter

The human body contains vast numbers of atoms, arranged in a highly organized way. An atom is made up of particles called **protons**, **neutrons**, and **electrons**. It is the smallest part of an element that can exist on its own, but in the body most atoms do not exist separately. Instead, they form parts of chemical compounds.

Carbon atom
In this carbon atom, six electrons (blue) surround the six protons (red) and six neutrons (gray) that form the nucleus.

Element

A pure substance that contains only one kind of atom

Just over 90 elements occur naturally on the Earth, but of these, only about 25 are found in the human body. Four elements – carbon, **hydrogen**, **oxygen**, and **nitrogen** – make up over 95 percent of the body's weight. We get these elements mainly by eating organic compounds in food and by drinking water. Many of the other 21 elements are needed only in tiny quantities. Most of these can be obtained by eating food that contains minerals ■.

Carbon

An element that is found in all living things

Life on Earth is based on the element carbon. A single carbon atom can form chemical bonds with up to four other atoms. These can be carbon atoms, atoms of different elements, or a combination of the two. This ability to link up with many other atoms allows carbon to form a vast range of chemical compounds.

Molecule

A chemical unit made up of two or more atoms joined together

A molecule is made of atoms held together by chemical bonds. The smallest molecules contain just two atoms, but the largest – such as those of DNA ■ – contain over a million. Giant molecules are known as **macromolecules**. Macromolecules containing carbon are an essential part of the human body and of all living things.

Cysteine molecule
Cysteine is one of the few amino acids to contain sulfur. It is an important ingredient in the structural protein keratin.

Chemical bond

A link between atoms

Chemical bonds hold compounds together. There are two main types. In a **covalent bond**, neighboring atoms share some of their outermost particle or electrons. Nearly all of the bonds in the body's carbon-containing molecules are covalent bonds. In an **ionic bond**, atoms lose or gain electrons instead of sharing them. This creates electrical forces that hold the atoms together. Ionic bonds are found in ionic compounds such as **salt** and other minerals.

Key to atoms

☐ Hydrogen
■ Carbon
■ Oxygen
☐ Sulfur
■ Nitrogen

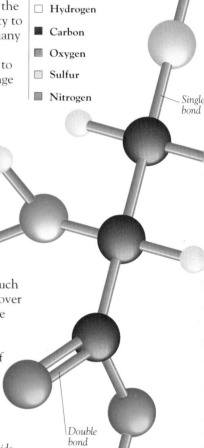

Single bond

Double bond

Chemical compound

A substance formed when two or more elements combine

The atoms in a compound are combined in exact proportions, and are linked by chemical bonds. Compounds often have very different properties to the elements they contain.

Organic compound

A chemical compound that contains carbon

The human body has four main types of organic compound – carbohydrates ■, proteins ■, lipids ■, and nucleic acids ■. These are all based on the element carbon, and often have large, complicated molecules.

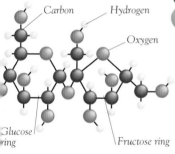

Carbon   Hydrogen

Oxygen

Glucose ring

Fructose ring

Sucrose molecule
Table sugar is made up of an organic compound called sucrose, which is a carbohydrate. A sucrose molecule consists of a six-sided glucose ring and a five-sided fructose ring.

Inorganic compound

A chemical compound that does not contain carbon

Most of the inorganic compounds in the body are simple substances. They include water, the body's most abundant inorganic compound, as well as the gas oxygen (O_2) and minerals. The gas **carbon dioxide** (CO_2) is usually classed as an inorganic compound, even though it contains carbon.

Ionic compound

A compound that contains ions

Table salt is an example of an ionic compound. It contains sodium atoms that have lost electrons, and chlorine atoms that have gained them. Atoms like this are known as **ions**. When you eat salt, the sodium and chlorine ions dissolve and separate. The body's fluids contain many different ions, and their levels are carefully regulated.

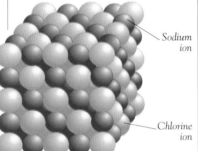

Sodium ion

Chlorine ion

Salt crystals
Table salt is an ionic compound called sodium chloride. It is made up of sodium ions and chlorine ions.

Chemical reaction

A chemical change

During a chemical reaction, elements or compounds undergo a change. Chemical reactions are an essential part of life, and many of them are accelerated by special proteins called enzymes ■. Some chemical reactions break down substances, and others build them up. Together, these reactions make up the body's metabolism ■.

Oxidation

A chemical reaction in which a substance combines with oxygen

The body uses oxygen to release energy in a process called aerobic respiration ■. Without a steady supply of oxygen, the human body cannot survive for more than a few minutes.

Water

A liquid that is essential to life

Drinking water replaces lost fluids

Water (H_2O) is a compound of hydrogen and oxygen. It is found in all living things, and it makes up between 50–75 percent of the body's weight. Water is important for several reasons. It is a very good solvent, and it can take part in many chemical reactions. It is also very good at holding heat, which helps the body to stay at a constant temperature. Finally, it flows easily, which helps to carry substances from place to place.

Solution

A uniform mixture of one substance dissolved in another

The body contains many dissolved substances, or **solutes**. In most cases, the substance in which they are dissolved – the **solvent** – is water. One important group of organic compounds, called lipids, does not dissolve in water. Lipids play a major part in separating cells ■ from their watery surroundings.

Concentration

The strength of a solution

If a solution is very strong, it is **concentrated** and contains a large quantity of solute compared to the quantity of solvent. If it is weak or **dilute**, it contains a small amount of solute.

See also

Aerobic respiration 104
Carbohydrate 22 • Cell 26 • DNA 25
Enzyme 24 • Lipid 23
Metabolism 102 • Mineral 108
Nucleic acid 25 • Protein 24

Carbohydrates

Carbohydrates are the body's main chemical fuel. Breaking them down in a controlled way releases the energy needed to power the body's cells.

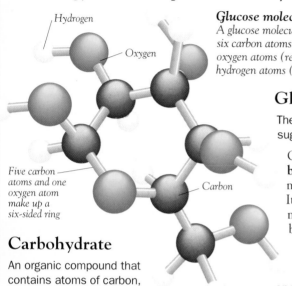

Hydrogen

Oxygen

Five carbon atoms and one oxygen atom make up a six-sided ring

Carbon

Glucose molecule
A glucose molecule is made up of six carbon atoms (black), six oxygen atoms (red), and 12 hydrogen atoms (white).

Carbohydrate

An organic compound that contains atoms of carbon, hydrogen, and oxygen

A carbohydrate molecule usually contains twice as many hydrogen ■ atoms ■ as oxygen ■ or carbon ■ atoms. The simplest carbohydrate is a **monosaccharide**. Its molecules ■ usually have six carbon atoms or less, and the central atoms are often arranged in a ring. A **disaccharide** is made up of two monosaccharides linked together; a **polysaccharide** is a long chain of tens or hundreds of monosaccharides.

Sugar

A simple carbohydrate that has a sweet taste

Most sugars are monosaccharides or disaccharides. They include glucose, which is found in most plant and animal cells ■, **fructose** or **fruit sugar**, which is found in fruit and honey, and **sucrose** or **cane sugar**, which is found in plant sap. Sugars give the body an almost instant energy supply.

Glucose

The most abundant sugar in the body

Glucose, or **blood sugar**, is a monosaccharide. It is the body's main fuel, and is broken down during aerobic respiration ■ to release energy. The body obtains glucose directly from food, or by digesting ■ other carbohydrates, such as starch or sucrose. The level of glucose in the blood is controlled by hormones ■. Glucose can be stored by converting it into glycogen, and released again by breaking down the glycogen.

Lactose

A disaccharide found in milk

Lactose is a sugar that provides energy for growing infants. It is found in human milk ■, and also in cow's milk and dairy products. Lactose is broken down during digestion by an enzyme ■ called **lactase**.

Milk
Some adults cannot make the enzyme lactase and so cannot digest the lactose found in milk.

Starch

A polysaccharide that plants use to store energy

Starch is found in cereals such as wheat and rice, and in other crops, including potatoes. Its molecules form long branching chains. During digestion, starch is broken down to form a sugar called **maltose**, which is made of two glucose units. Maltose is then broken down into glucose itself.

Starch molecul

Cellulose

A polysaccharide that plants use as a building material

Cellulose is similar to starch. It is found in vegetables and other plant-based foods. Unlike starch, cellulose cannot be digested. It passes straight through the body, and forms a major part of dietary fiber ■.

Glycogen

A polysaccharide that the body uses as an energy store

Glycogen is also called **animal starch**. The body makes glycogen by linking glucose molecules together. Glycogen is stored in the liver ■ and in muscles ■. When the body runs short of glucose, glycogen is broken down, and the glucose that is formed is released into the blood.

Lipids

Unlike most other organic compounds, lipids usually repel water. They are important building materials in cells because they can partition things from their watery surroundings. Lipids are also a concentrated source of energy.

Lipid

An insoluble organic compound

Lipids include fats and oils, together with other fat- and oil-based substances. Their molecules ■ contain carbon and hydrogen atoms, together with a few oxygen atoms. The body uses lipids in many ways. Some of them form the plasma membranes ■ around cells, or the membranes around organelles ■. Others act as an energy store, as a protective cushion, or as insulation against the body losing too much heat. A few act as chemical messengers, or hormones ■.

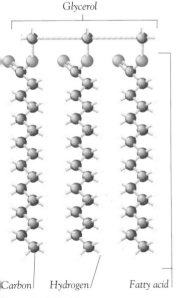

Glycerol

Carbon   Hydrogen   Fatty acid

A typical lipid
Lipids such as fats and oils contain complex molecules called triglycerides, which store energy. Triglycerides are formed by fatty acids and glycerol.

Fat

A lipid that is normally solid at room temperature

Fats consist of chemical compounds called **triglycerides**. Each triglyceride molecule contains a substance called **glycerol**. This is linked to three **fatty acids**, which are made almost entirely of carbon and hydrogen atoms arranged in a long chain. A **monoglyceride** contains glycerol with one fatty acid. In a **saturated fat**, the carbon atoms are held together by single chemical bonds ■. Saturated fats are common in the body, and are also found in meat, eggs, and dairy products.

Oil

A lipid that is normally liquid at room temperature

Oils are liquid fats. They are common in plants, and they make up an important part of a balanced diet ■. Many plant oils are **unsaturated**, which means that their carbon chains contain at least one double chemical bond. **Polyunsaturated fats**, such as sunflower oil, contain more than one double bond.

Steroid

A complex lipid in which the carbon atoms are arranged in rings

Steroids are varied substances that carry out a range of tasks. They include several hormones, some vitamins ■, and cholesterol.

Cholesterol

A steroid used to make hormones and transport fats

Cholesterol is needed for the formation of plasma membranes, steroid hormones, and bile salts ■. It is made by the liver ■, and is also found in meat, eggs, and dairy products. Although the body needs cholesterol, excess amounts are linked to a dangerous disease called atherosclerosis ■.

Phospholipid

A form of lipid that makes up membranes

Phospholipid molecules make up the plasma membranes around cells, and also the membranes around some organelles. Each phospholipid molecule has a "head" and two "tails." The head is attracted by water, or **hydrophilic**, while the "tails" are repelled by it, or **hydrophobic**. In the body, the most important phospholipid is **lecithin**, or **phosphatidyl choline**. Lecithin is also found in egg yolks.

Egg yolk contains lecithin

Oil on water
An oily substance forms a thin film when it spreads over water.

Proteins

Proteins are among the most complicated substances in the body. There are thousands of different kinds, and they perform many different tasks – from building structures such as hair and nails to controlling chemical reactions.

Protein

An organic compound made up of amino acid units

A protein is an organic compound ■. Its molecules ■ are made up of chains of amino acid units linked together in an exact sequence. During protein synthesis ■, the chain folds up, which gives the finished molecule a characteristic shape. The precise shape of a protein molecule enables it to do a particular job. The body makes proteins by following the coded instructions held in genes ■.

Amino acid

A protein building block

There are 20 different types of amino acid in the body, so they can be arranged in many different sequences. Amino acids contain carbon ■, oxygen ■, and hydrogen. They also have at least one **amino group** (NH_2), which has one nitrogen atom and two hydrogen atoms. The chemical bond between neighboring amino acids is called a **peptide linkage**, or **peptide bond**, and the chain is called a **peptide**.

Essential amino acid

An amino acid that the body cannot make

The human body cannot make amino acids from simple raw materials, but it can convert some amino acids into others. Essential amino acids make up 10 of the 20 amino acids that the body needs, and have to be obtained from food. The remaining 10 amino acids can be made from these essential amino acids by chemical conversion.

Enzyme

A kind of protein that speeds up chemical reactions in the body

Enzymes act as **catalysts**, which means they speed up chemical reactions ■ thousands or millions of times. Without enzymes, chemical reactions such as digestion would occur so slowly that life would be impossible.

An enzyme-controlled reaction
Each of the body's enzymes speeds up a single reaction. The molecules involved fit onto the enzyme like a key in a lock.

Structural protein

A protein that is used as a building material

Structural proteins support and protect many parts of the body. The body's most abundant structural protein is **collagen**. It forms tough fibers that strengthen cartilage ■, tendons ■, and other kinds of connective tissue ■ such as bone ■. Skin, hair, and nails contain the structural protein **keratin**. Keratin is tough, waterproof, and able to resist chemical attack. Another protein, **elastin**, makes skin taut and allows arteries ■ to expand and contract as the heart beats.

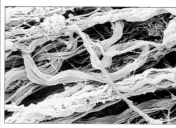

Collagen fibers
Collagen is made up of tough, flexible fibers that stop tissues from being torn.

Transport protein

A protein that is used to carry other substances

Transport proteins collect substances from one place and release them in another. The most important transport protein is hemoglobin ■. It carries oxygen in the blood.

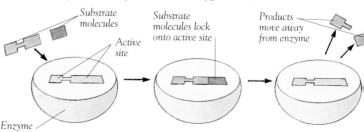

Substrate molecules

Substrate molecules lock onto active site

Active site

Products move away from enzyme

Enzyme

1 *The molecules involved in a reaction, called substrates, fit into a part of an enzyme called the active site.*

2 *The enzyme holds the substrates together. The substrates react and form products.*

3 *The products are released. The enzyme is unaffected by the reaction, and is ready to attract more molecules.*

Nucleic acids

Nucleic acids are the body's equivalent of a computer program. Using a special chemical code, they store all the information needed to assemble the entire body and to make it work.

Nucleic acid

A complex organic compound that carries information

Nucleic acids are organic compounds ■ made up of large numbers of small units called nucleotides. There are two kinds of nucleic acid in the body – DNA and RNA. DNA carries the body's genes ■. Together, these genes make up a "plan" of chemical information that controls the body by telling cells how to make proteins ■. DNA is found mainly in cell nuclei ■. RNA acts as a "shuttle" service. It copies the information in DNA and carries it away from the nucleus to be put into action.

DNA

A substance that stores information in the body

A molecule of DNA, or deoxyribonucleic acid, consists of millions of atoms. It is formed by two slender strands that spiral around each other to form a shape called a double helix. The strands are held together by chemicals called bases. The exact sequence of bases forms all the genetic instructions in each cell, and varies slightly from one person to another. Each of the body's cells contains a complete set of DNA molecules, packaged into threadlike structures called chromosomes ■.

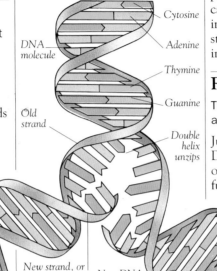

DNA molecule
Cytosine
Adenine
Thymine
Guanine
Old strand
Double helix unzips
New strand, or complementary copy
New DNA molecule

Replicating DNA
During replication, the double helix in a DNA molecule "unzips," and its strands separate. The two strands then form complementary copies of themselves. With these copies, two new DNA molecules are produced, each with one old strand and one new strand.

Base

A substance that forms a store of information in a nucleic acid

Each kind of nucleic acid has just four bases. Bases work like the letters in the alphabet, and spell out a list of instructions that are written in genetic code ■. When they are decoded, these instructions tell a cell how to make all the proteins that it needs. In DNA, the bases are cytosine, guanine, thymine, and adenine. RNA has the same bases, but uses uracil instead of thymine.

Base pair

Two bases linked chemically

In a DNA molecule, every base is chemically linked to its partner on the opposite strand. Adenine always pairs with thymine, and cytosine always pairs with guanine. This precise pairing means that each strand carries the same information, but in a different form. When the strands copy themselves, the information is passed on.

Replication

The process by which a nucleic acid molecule copies itself

Just before a cell divides ■, its DNA molecules copy themselves, or replicate. This ensures that a full set of instructions is passed on to each new cell.

RNA

A nucleic acid that handles the information used to make proteins

There are two main kinds of RNA, or ribonucleic acid. Messenger RNA, or mRNA, copies part of the information held in a DNA strand. It then carries it out of the nucleus so that it can be used to make a protein. This process is called transcription ■. Transfer RNA, or tRNA, collects the amino acids needed to make the protein. It delivers them to a ribosome ■, which "reads" the RNA message and links the amino acids in the correct order. This process is called translation ■.

See also

Cells

Our bodies contain over 50,000 billion cells that work closely together. Each one is a microscopic world in which thousands of chemical processes are carried out in an organized way.

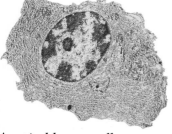

A typical human cell
The central nucleus is surrounded by folded layers of endoplasmic reticulum.

Cell

A tiny unit of living matter

Cells are the smallest parts of the body that are completely alive. A typical cell consists of a thick fluid called cytoplasm, which contains a nucleus and is surrounded by a thin barrier called a plasma membrane ■. This allows selected substances into or out of the cell. There are many different kinds of cells in the body, and they carry out a wide variety of tasks. Instead of being mixed up at random, they are usually arranged into groups called tissues ■. Cells reproduce by dividing ■ in two.

See also

Organelle

A structure inside a cell that has a particular function

If you think of a cell as a factory, organelles are its separate departments. Each one carries out a particular task in maintaining the life of a cell. Organelles keep the many chemical processes of a cell separate, so that different chemical reactions do not interfere with one another. The number and type of organelles vary according to the function of the cell.

Cytoplasm

The contents of a cell, excluding the plasma membrane and the nucleus

Cytoplasm is a transparent, jellylike fluid that contains organelles. It often circulates inside the cell. Up to 90 per cent of cytoplasm is water, which contains dissolved ions ■, microscopic particles, and organic compounds ■.

Cytoskeleton

A network of fine chemical threads throughout a cell

Cytoplasm contains a network of extremely small threads and tubes, visible only under a very powerful microscope ■. These make up a cell's chemical "skeleton." The threads keep the organelles within a cell in the correct position, and move them during cell division. They also allow a cell to change shape.

Endoplasmic reticulum

A system of membranes that make and stores a range of substances

The endoplasmic reticulum, or **ER**, is a continuous network of folded membranes that acts like a large work surface inside a cell. There are two kinds of ER. **Rough endoplasmic reticulum** is covered with small organelles called ribosomes, which make proteins ■. **Smooth endoplasmic reticulum** does not have ribosomes, and is involved in making lipids ■.

Structure of a cell
Cells are packed with many different organelles and a large nucleus.

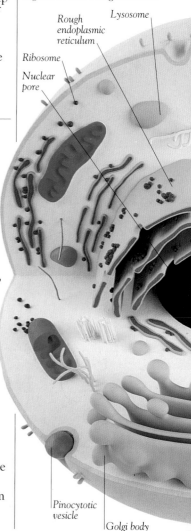

Rough endoplasmic reticulum

Lysosome

Ribosome

Nuclear pore

Pinocytotic vesicle

Golgi body

Ribosome

A cluster of chemicals that assembles proteins

Ribosomes are among the smallest organelles in a cell, but they are often very numerous. They can be scattered in the cytoplasm, or attached to the endoplasmic reticulum. Ribosomes play a key part in protein synthesis ■ by translating ■ the instructions carried by messenger RNA ■. Cells that make a lot of protein have large numbers of ribosomes, which gives their cytoplasm a very speckled appearance.

Nucleus
Nuclear membrane
Mitochondrion
Mitochondrial cristae
Smooth endoplasmic reticulum
Cell membrane
Cytoplasm

Golgi body

A packaging and transport system for substances made by a cell

Many of the body's cells make special substances that are released onto their surface or into their surroundings. The Golgi body is made up of membranes that store these substances, and then eject them from a cell. The Golgi body also makes lysosomes.

Nucleus

The control center of a cell

The nucleus (plural **nuclei**) is the largest organelle in many cells, and often the only one that can be seen with an ordinary microscope. It houses a cell's chemical information, which is stored in molecules of DNA ■. The DNA molecules are wound up to form threadlike structures called chromosomes ■. The nucleus is separated from the rest of the cell by a **nuclear envelope**, or **nuclear membrane**. This contains holes called **nuclear pores** that allow chemical contact between the nucleus and its surroundings. The term nucleus comes from a word that means "nut."

Lysosome

A reservoir of digestive chemicals

Lysosomes contain powerful enzymes ■ that can digest living matter. They break down worn-out organelles, and digest foreign substances that a cell has engulfed during phagocytosis ■. Sometimes, lysosomes destroy the cell that surrounds them. This is called "self-digestion," or **autolysis**, and it is common during the early stages of development ■.

Mitochondrion

An organelle that releases energy

Mitochondria (the plural of mitochondrion) carry out aerobic respiration ■ to release energy. They have two membranes: the outer membrane is smooth, but the inner one has folds called **cristae**. This is where respiration takes place. If a cell needs a lot of energy, its mitochondria can grow and divide to meet this need.

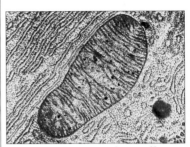

Mitochondrion inside a cell
This oval-shaped mitochondrion has a heavily folded inner membrane.

Cilium

A short, hairlike projection that sweeps backward and forward

Cells use **cilia** (plural of cilium) to move things across their surface. Cilia are found in the Fallopian tubes ■ and the respiratory tract ■.

Flagellum

A long, hairlike projection that is mainly used in movement

A flagellum (plural **flagella**) is usually single and lashes from side to side to move a cell along. Flagella are found in sperm ■ cells.

Extracellular material

A nonliving substance found outside cells

Most of the body's cells are bathed in tissue fluid ■. Some cells are surrounded by solid substances as well, including proteins such as keratin ■, collagen ■, and elastin ■, and also some minerals ■.

Cell membranes

Cell membranes can only be seen by the most powerful microscopes. They act like a protective shield, allowing some things to travel into and out of cells, but preventing the passage of others. Despite their fragile appearance, cell membranes are extremely resilient, and automatically seal themselves if they are broken.

Plasma membrane

A thin barrier that separates a cell from its surroundings

A **membrane** is any thin barrier around an organelle ■, a cell ■, or a part of the body. A plasma membrane, or **cell membrane**, surrounds a cell, and is made of a double layer of phospholipid ■ molecules ■. The molecules fit together like pieces in a moving mosaic, and they quickly seal any small gaps that develop. Plasma membranes also contain other lipids ■, such as cholesterol ■, and a number of proteins ■. Many of the proteins act as channels to let substances pass into or out of the cell.

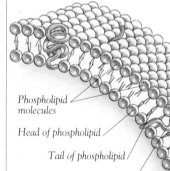

Phospholipid molecules

Head of phospholipid

Tail of phospholipid

Protein molecule within membrane

Cell membrane structure
Cell membranes are made of a double layer of phospholipid molecules. The heads of the phospholipids are attracted to the watery surroundings of a cell; the tails are repelled by water. This makes the molecules line up in a double layer.

Differentially permeable membrane

A porous barrier

A **permeable membrane** is one that lets molecules pass through it. The plasma membranes around cells are differentially permeable, or **semipermeable**, which means that they let some molecules pass through, but block the path of others. This exchange of chemicals between cells and their surroundings allows cells to stay alive. Unlike a nonliving membrane, a plasma membrane is influenced by chemicals such as hormones ■ and histamines ■. These can change the membrane's permeability, altering the chemical balance of the cell.

Diffusion

The spreading out of a substance from an area of high concentration

Diffusion occurs in the body and in nonliving matter. It does not require an input of energy. During diffusion, molecules move apart until they are evenly spread out. **Facilitated diffusion** occurs when substances such as glucose ■ are "helped" through a cell's plasma membrane by special proteins. As with normal diffusion, this process does not require an input of energy.

Osmosis

The flow of water through a semipermeable membrane from a weak to a strong solution

Osmosis is a form of diffusion that takes place when two solutions ■ of different strengths or concentrations ■ are separated by a semipermeable membrane. The membrane stops large dissolved molecules from passing through, but allows the passage of water molecules. The water molecules move through the membrane until the solutions are of equal strength. Osmosis occurs in the body and in nonliving matter. It does not require any energy. If water moves out of a cell faster than it enters, the cells shrink and may become bumpy. This process is called **crenation**.

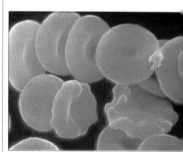

Crenation of red blood cells
As water moves out of red blood cells, the cells shrink and become notched.

Active transport

A form of transport that uses energy to move molecules across a membrane

Unlike diffusion and osmosis, active transport makes solutions more concentrated. It requires energy, which usually comes from a substance called ATP ■. Cells use active transport to collect chemicals from their surroundings or to pump them out. This process often uses protein molecules to carry a substance through a membrane. The sodium-potassium pump ■ in nerve cells ■ works by active transport.

Key
● Solute molecule □ Outside cell
○ Water molecule □ Inside cell
▪ Cell membrane

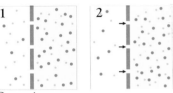

Diffusion
1) There is a higher concentration of solute molecules outside the cell, so (2) solute molecules pass from an area of high to low concentration, thereby evening out the difference.

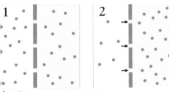

Osmosis
1) There is a lower concentration of water molecules inside the cell, so (2) water molecules travel through the cell membrane, which evens out the strength.

Active transport
1) The concentration of solute molecules inside and outside the cell is equal. (2) Solute molecules are actively transported across the membrane, which results in solute molecules being more concentrated inside the cell than outside.

Osmotic pressure

The pressure that stops water from moving into a solution through a semipermeable membrane

As water moves during osmosis, it pushes against its surroundings, creating an increase in pressure. The osmotic pressure of a solution is greater when there is a higher number of solute particles in a given volume of solution. This causes more water molecules to move into the solution.

Isotonic

Having the same osmotic pressure

The osmotic pressure of cells and body fluids is continuously monitored and adjusted. This process is called osmoregulation ▪. The body's cells are normally isotonic with their surroundings, which means that they do not gain or lose water through osmosis. If a cell becomes **hypotonic** to its surroundings, its fluid is less concentrated than that around it. Osmosis will then move water out of the cell, so the cell will shrink. If a cell becomes **hypertonic**, its fluid is more concentrated than its surroundings. In this case, osmosis will move water into the cell, so the cell will swell up.

Water potential

The tendency of water to move through a differentially permeable membrane during osmosis

A cell's water potential depends on the osmotic pressure of the fluid inside it. If a cell contains a high concentration of dissolved substances, and therefore a high osmotic pressure, its water potential will be low. The higher water potential outside will drive water into the cell.

Pinocytosis

The engulfing of liquid by a cell

During pinocytosis, a cell surrounds a droplet of liquid and engulfs it. The cell membrane folds inwards to form a pocket called a **vesicle**. The liquid is held in the vesicle, where it is digested. Pinocytosis is carried out by most of the body's cells, whereas only specialized cells can carry out phagocytosis.

Phagocytosis

The engulfing of particles by a cell

As well as taking in individual molecules, many cells also absorb particles or complete cells from their surroundings. They do this to clear away debris or to attack invaders that cause disease ▪. During phagocytosis, a cell produces slender outgrowths called **pseudopodia** (singular **pseudopodium**). These surround the particle or cell, and join together to engulf it. The engulfed material is stored in a membrane-lined pocket called a vesicle. The vesicle then joins up with one or more lysosomes ▪, and its contents are digested. Cells that specialize in phagocytosis are called phagocytes ▪.

Phagocytosis
This cell has engulfed a large number of bacteria, which are the green and red structures inside the cell.

See also

ATP 105 • Cell 26 • Cholesterol 23
Concentration 21 • Disease 92
Glucose 22 • Histamine 94
Hormone 78 • Lipid 23
Lysosome 27 • Molecule 20
Nerve cell 58 • Organelle 26
Osmoregulation 77 • Phagocyte 95
Phospholipid 23 • Protein 24
Sodium-potassium pump 59
Solution 21

Cell division

Every day, your body produces billions of new cells by cell division. Some of these new cells are needed for growth, while others replace those that have died. A special kind of cell division makes the cells that are used for reproduction.

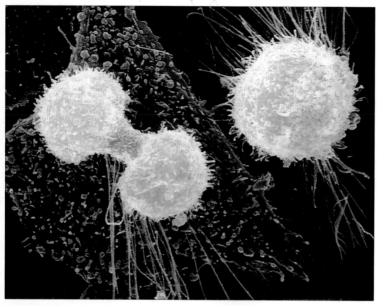

Cell division
This false color electron micrograph shows the last phase of mitosis. The two new cells, left, are still connected by the cytoplasm. The cytoplasm then divides and the two new cells are complete.

Cell division

The process by which cells reproduce

Cells divide in two different ways – mitosis and meiosis. Mitosis is used for growth and repair, and involves a single division that produces two new cells. Some cells continue to divide, while others become adapted, or **specialized**, for a particular kind of work, and lose the ability to divide. Meiosis is used for reproduction. It involves two divisions that create four new cells.

Cell cycle

The life cycle of a cell

Like the body as a whole, each cell has its own life cycle. The length of this cycle varies from one kind of cell to another. Cells that experience a lot of wear and tear, such as those lining the skin ▪ and alimentary canal ▪, are produced by cell division about every 24 hours. Highly specialized cells often divide much less frequently than this. For example, neurons ▪, or nerve cells, do not divide at all once they have been formed. Division actually takes about one hour. During the remaining period, called **interphase**, each cell carries out its normal work, and copies its chromosomes ▪ in preparation for the next division.

Somatic cell

A cell that is not involved in sexual reproduction

Somatic cells make up the vast majority of cells in the body. They are produced by mitosis. Although somatic cells specialize in different tasks and often look different, they have identical genes ▪. They are also diploid ▪, which means that they contain a double set of chromosomes. Somatic cells are not involved in sexual reproduction, and therefore cannot pass on genes to the next generation.

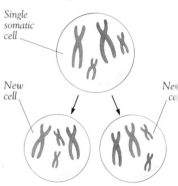

Single somatic cell

New cell

New cell

Mitosis
Mitosis involves a single division and produces two new cells. Each cell has a double set of chromosomes and is genetically identical to the parent cell.

Mitosis

The division of a cell nucleus to produce two identical cells

Mitosis is the process that the body uses for growth and repair. During mitosis, a single somatic cell produces two new cells with identical chromosomes. Just before mitosis begins, each of the cell's chromosomes copies itself and forms two X-shaped structures called chromatids ▪. During mitosis, the chromatids tighten up, or condense. They are then pulled apart and two new nuclei ▪ are formed. After mitosis, the cytoplasm divides and the two new cells are complete.

Meiosis

A form of cell division that produces differing cells

Meiosis takes place in the reproductive system ■, and produces sex cells ■. It involves two cell divisions, one after the other, and results in four new cells. Each new cell is haploid ■, which means it has a single set of chromosomes and is genetically unique. In males, meiosis is part of spermatogenesis ■, which occurs in the testes ■. In females, meiosis is part of oogenesis ■, which occurs in the ovaries ■.

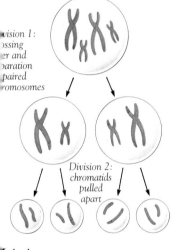

Division 1:
crossing
over and
separation
of paired
chromosomes

Division 2:
chromatids
pulled
apart

Meiosis
During the two divisions of meiosis, the original cell's genes are mixed and recombined. The four new cells that are produced are genetically unique.

Spindle

A network of tiny tubes that separates chromatids when a cell divides

During cell division, chromatids separate and end up in different cells. A spindle is the structure inside a cell that pulls them apart. It has two poles, formed by organelles ■ called **centrioles**. The poles are connected by tiny tubes, and these drag the chromatids toward them.

Crossing over

The swapping of genes between matching chromosomes

At the beginning of meiosis, pairs of matching or homologous chromosomes ■ line up alongside each other and form structures called **tetrads**. Each chromosome has two chromatids, so a tetrad is a single unit of four chromatids (tetrad comes from the Greek word for four). The chromatids in each tetrad form bridges called **chiasmata** (singular **chiasma**), and swap pieces with each other. Afterwards, the chromosomes pull apart. Crossing over is an important process, because it is one way in which the body brings about a new combination of genes, or **recombination**.

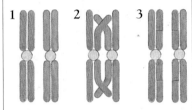

1   2   3

Recombination
(1) Before meiosis occurs, matching chromosomes line up to form tetrads. (2) The chromatids that make up the chromosomes cross over and swap pieces. (3) The chromosomes separate. Each pair now has a new combination of genes or alleles from each parent.

Random assortment

The random division of a double set of chromosomes

Meiosis begins with a single cell that has a double set of chromosomes. One set originally comes from the father, and one from the mother. During meiosis, the set is divided, and the new cells usually receive a mixture of chromosomes from both parents. This is called random assortment, and it is another way in which the body brings about recombination.

See also

Alimentary canal 116 • Chromatid 132
Chromosome 132 • Diploid cell 133
Gene 132 • Haploid cell 133
Homologous chromosome 133
Neuron 58 • Nucleus 27
Oogenesis 129 • Organelle 26
Ovary 129 • Reproductive system 128
Sex cell 128 • Skin 32
Spermatogenesis 128
Testis 128 • Virus 92

Cancer

A disease produced by uncontrolled cell division

There are many different kinds of cancer. The one feature they both have is that they involve uncontrolled cell division, which leads to growths called **tumors**. Once a tumor has formed, it often expands rapidly, and it may stop part of the body from functioning normally. Tumors often **metastasize**, which means that some of their cells break away and start new tumors in other places. Scientists believe that cancers start when cancer-producing genes, which exist in normal cells, are triggered into action. The factors that trigger these genes are very varied. They include chemicals called **carcinogens**, accidental rearrangements in a cell's DNA, and attack by viruses ■.

Cancer cells
This light micrograph shows squamous carcinoma cells, which are responsible for skin cancer.

Skin, hair & nails

The skin is the largest organ in the body. Together with the hair and nails, it forms a highly effective barrier between the body and the outside world.

Dermis

The inner part of the skin

The dermis is thicker than the epidermis and contains only living cells. It contains a dense network of blood capillaries ■, hair follicles, and sweat glands ■. There are also nerve endings and receptors ■ that sense pressure, temperature, and pain. Fibers of collagen ■ and elastin ■ within the dermis allow it to stretch and then return to its normal shape.

Skin structure
The skin has many layers which have various structures embedded within them

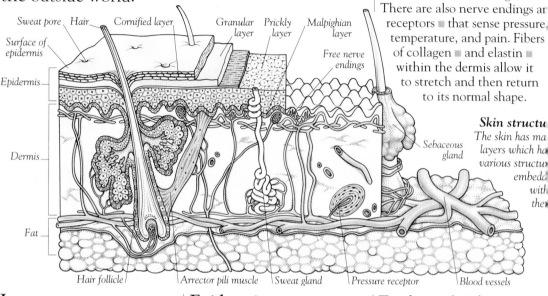

Sweat pore · Hair · Cornified layer · Granular layer · Prickly layer · Malpighian layer

Surface of epidermis

Free nerve endings

Epidermis

Sebaceous gland

Dermis

Fat

Hair follicle · Arrector pili muscle · Sweat gland · Pressure receptor · Blood vessels

Integumentary system

A collection of organs and structures that protect the body's exterior

An **integument** is a kind of outer covering. The integumentary system consists of the skin and structures formed by it, such as hair and nails. It protects the body and plays a part in homeostasis ■.

Skin

The outer body covering

Skin is an organ ■ that consists of several layers of cells ■. It forms a waterproof barrier that stops the body from drying out. It also protects it from physical damage, from microorganisms ■ that could cause infection ■, and from ultraviolet rays in sunlight. Skin plays a part in thermoregulation ■, and contains nerve cells that detect pressure, temperature, and pain. In an adult, it has a surface area of up to 21.5 ft² (2 m²).

Epidermis

The outer part of the skin

The epidermis is a thin sheet that contains at least four different layers of cells. The deepest of these is the **Malpighian layer**, or **basal layer**. This contains a single row of cells that continuously divide, with the new cells being pushed outward. Above the Malpighian layer is the **prickly layer**, or **stratum spinosum**, which consists of 8–10 rows of cells covered in spinelike projections. As the cells continue to push outward, they enter the **granular layer**. Here, their nuclei ■ start to break down, and they become filled with keratin ■. By the time the cells reach the outermost layer, called the **cornified layer**, they are flat and completely dead. The outermost cells are constantly worn away, and are replaced by new cells growing up from below.

Epidermal ridge

A ridge in the skin of the hands and feet

Hands and feet spend a lot of their time gripping or pushing against the ground. To prevent them slipping, their working surfaces are covered with densely packed ridges less than 0.04 inch (1 mm) high. The ridges are arranged in distinctive patterns that vary from one person to another. They contain sweat ducts, and they may leave behind a mark, or **fingerprint**, when an object or surface is touched.

Fingerprints
With the exception of identical twins, every person has unique fingerprints.

Melanin

A pigment found in the skin

Melanin is a brownish black substance found in the epidermis of the skin, in hair, and also in the retina ■ and iris ■ of the eye. It absorbs harmful ultraviolet waves in sunlight, and prevents them from damaging the body. Exposure to ultraviolet light increases the production of melanin and darkens, or tans, the skin. People with dark skin have a lot of melanin in their skin. People with light skin have only a little, although sometimes it may be concentrated in small patches of darker skin called **freckles**. Melanin production is controlled by a single gene ■. People who lack this gene are called **albinos** ■, and have very pale skin and white or pale yellow hair.

Melanin
The color of a person's skin is determined by the amount of melanin their skin produces. There is a wide variation in skin color between people of different racial groups.

Sebaceous gland

A gland that produces oil

Sebaceous glands are usually connected to hair follicles. They produce an oily liquid called **sebum**, which keeps skin and hair soft and flexible. Sebaceous glands are found over the whole body, except the palms of the hands and soles of the feet. They are quite small in the skin of the arms, legs, and trunk, but are larger and more numerous in the face and neck.

Sweat gland

A coiled gland that produces sweat

Sweat glands lie in the dermis. Each one is connected to a small depression in the surface, called a **sweat pore**, by a short **sweat duct**. When the body overheats, the glands produce a salty fluid called **sweat**. This oozes onto the skin and evaporates. As it changes into a vapor, it absorbs heat from blood flowing below the skin's surface and cools the body. The process of producing sweat is called perspiration ■.

Sweat pore

Hair

A filament of dead cells rooted in the skin

Hair grows over the entire body, except on the palms of the hands, the soles of the feet, the lips, parts of the genitals, and the nipples. It protects the skin and helps us to feel things that come near the skin's surface. Each hair is made of dead cells containing keratin. Hair texture depends on where and when it grows. A developing fetus ■ is covered with fine hair called **lanugo**, but this is shed before birth. Lanugo is replaced by soft **vellus hair**, which grows over the body, and by thicker **terminal hair**, which grows on the scalp and forms eyebrows and eyelashes. Hairs grow out of **hair follicles,** hollow spaces in the skin called that reach through the epidermis and dermis. Cells at the base of the follicle divide to form the hair, and make it grow by up to 0.4 inch (1 cm) a month. Follicles with a round opening produce straight hair. Follicles with an oval or curved opening produce curly or wavy hair.

Goose pimple

A raised bump surrounding a hair follicle

When you get cold or are suddenly frightened, your skin may become covered in goose pimples, or **goose bumps**. Each bump is formed by a small smooth muscle ■ called an **erector pili**. These muscles pull your hair follicles upright, and make your hair stand on end.

Nail

A hard covering that protects a finger or toe

Nails protect fingers and toes from damage and allow small objects to be handled. They are made almost entirely of keratin. Nails are formed by cells that divide constantly, and they grow at a rate of up to 0.2 inch (5 mm) a month. The flap of skin at the growing base of a nail, called a **cuticle**, is part of the skin's cornified layer. The curved white area beneath the nail, called a **lunula**, contains some of the cells that divide to form the nail.

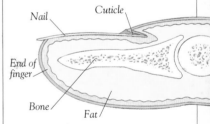

Structure of a nail
Nails grow from a layer of active cells under the skin at their base and sides.

See also

Albinism 146 • Capillary 88 • Cell 26
Collagen 24 • Elastin 24 • Fetus 137
Gene 132 • Gland 78 • Homeostasis 76
Infection 92 • Iris 68 • Keratin 24
Microorganism 92 • Nucleus 27
Organ 18 • Perspiration 77 • Receptor 59
Retina 69 • Smooth muscle 49
Thermoregulation 76

Tissues of the skeleton

The skeleton is the strong, flexible structure that supports your body. It provides your body with shape and protection and enables you to move. The skeleton is a living part of your body, adapting to the demands of everyday life, and repairing itself if damaged.

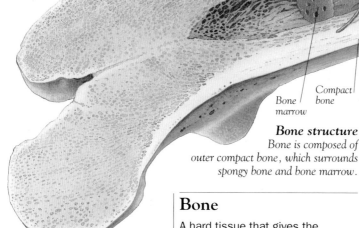

Bone marrow / *Compact bone*

Bone structure
Bone is composed of outer compact bone, which surrounds spongy bone and bone marrow.

Medullary cavity

Osteon • *Periosteum* • *Artery* • *Vein*

Spongy bone

Skeleton

A strong internal framework that supports and protects the body

The skeleton consists of the axial ■ and appendicular skeletons ■, which are made mainly of bone and cartilage. It gives the body its shape, protects internal organs, and provides an anchor for muscles, which allow the body to move. It also stores minerals ■ and forms blood cells.

See also

Appendicular skeleton 40
Axial skeleton 36 • Blood vessel 88
Cell 26 • Collagen 24
Connective tissue 19 • Embryo 137
Enamel 118 • Epiglottis 111 • Hormone 78
Intervertebral disc 37 • Joint 44
Membrane 28 • Mineral 108 • Nerve 58

Bone

A hard tissue that gives the skeleton its strength

Bone is a living connective tissue ■ composed of collagen ■, and the minerals calcium and phosphorus. It consists of widely spaced cells ■ called **osteocytes**, surrounded by fibers of collagen, which give the bone its strength. Bone is constantly built up by cells called **osteoblasts** and broken down by cells called **osteoclasts**.

Compact bone

A hard material that forms the exterior of bones

Compact bone is also called **hard bone, cortical or ivory bone**. It is the second hardest material in the body after enamel ■, and can bear powerful forces without breaking. Compact bone is very dense and is made up of parallel structures called osteons. These are visible under a microscope.

Osteon

A unit of compact bone

An osteon, or **Haversian system**, is a collection of tiny bony tubes called **lamellae** (singular **lamella**). These are arranged in circular layers around blood vessels ■ and nerves ■ that run down the center of the osteon, in a space called a **Haversian canal**. Mature bone cells, or osteocytes, live in spaces called **lacunae** between lamellae. The osteocytes are connected by microscopic channels.

Spongy bone

A honeycombed material in bones

Spongy, or **cancellous bone**, is made up of a honeycomb of bony struts called **trabeculae** (singular **trabecula**) and does not contain osteons. It is less dense than compact bone, which helps to reduce the skeleton's weight. The spaces inside spongy bone are often filled with red bone marrow.

Bone marrow

A soft tissue inside bones

Bone marrow is found within spongy bone and in the central space, or **medullary cavity**, of long bones. **Red marrow** forms blood cells; **yellow marrow** stores fat.

Periosteum

A membrane that covers the surface of bones

The periosteum is a thin but tough membrane ■. It covers the entire outer surface of a bone, except at a joint ■. The membrane forms an attachment point for tendons. It is essential for bone growth and repair, and contains capillaries that nourish bone.

Cartilage

A tough, flexible tissue that supports the body and eases movement

Cartilage is a kind of connective tissue formed by cells called **chondrocytes**. It contains fibers of the protein collagen, set in a jellylike substance. The collagen gives cartilage its strength, while the jelly allows it to be flexible. Cartilage contains few blood vessels and no nerves.

Hyaline cartilage

A type of cartilage that has a shiny appearance and slippery texture

Hyaline cartilage, or **gristle**, is the most common kind of cartilage. Hyaline means "glasslike," and this smooth, glossy cartilage lines many joints. It enables one bone to slide smoothly over another.

Fibrous cartilage

A type of cartilage that contains a large amount of collagen

Fibrous cartilage acts as a shock absorber. It is found in intervertebral discs ■, and also in the knees and wrists.

Elastic cartilage

A type of cartilage that contains strong fibers that can stretch

Elastic cartilage contains a springy protein called elastin. It is found in the epiglottis ■, and in the outer part of the ear.

Fracture

A break in a bone

A bone put under a lot of pressure may break. The bone will usually repair itself, but treatment may be needed to ensure that it returns to its original shape. The broken surfaces are often lined up and held together until the repair is complete. There are two main types of fracture. In an **open** or **compound fracture**, the broken bone projects through the skin; in a **simple** or **closed fracture**, it does not.

Repair of fractured bone
A broken bone goes through several stages before it is completely mended.

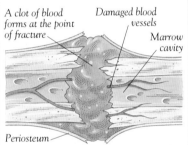

A clot of blood forms at the point of fracture

Damaged blood vessels

Marrow cavity

Periosteum

1 *A blood clot forms at the fracture site 6–8 hours after the injury.*

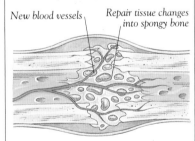

New blood vessels

Repair tissue changes into spongy bone

2 *Blood capillaries grow into the blood clot and damaged tissue is removed. The broken ends of the bone are connected by collagen fibers. Repair tissue called a callus forms. This is later replaced by spongy bone.*

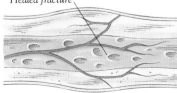

Healed fracture

3 *Remodeling occurs: compact bone replaces spongy bone at the edge of the fracture.*

Clopton Havers

English doctor (1650–1701)

People have studied bones for centuries, but Clopton Havers was one of the first to use a microscope to study their fine structure. He discovered that compact bone is made up of circular layers that surround tiny tubes. These are named Haversian systems. He wrote an important medical textbook called *Osteologia Nova*, which means "a new study of bones."

Ossification

The process of bone formation

When the skeleton first forms, it consists mainly of cartilage. Ossification starts when an embryo ■ is 6 weeks old and continues into adulthood. During ossification, osteoblasts deposit mineral salts in the cartilage, turning it into bone. Changes in hormone ■ levels later in life can make the skeleton lighter and weaker, so bones may break more easily. This is called **osteoporosis** and mainly affects elderly people.

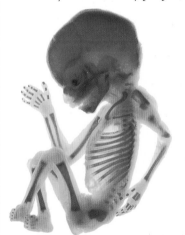

The fetal skeleton
This X-ray shows bone development in a 12-week-old fetus. Skeletal tissue that has developed into bone is stained red, while cartilage remains white.

Axial skeleton

The axial skeleton runs down the center of the body. It contains fewer than half of the body's 206 bones, and protects and supports important internal organs such as the heart and lungs. The axial skeleton also links with the bony girdles that anchor the limbs.

Axial skeleton

The central part of the skeleton

An imaginary line called the axis ■, or midline, runs down the middle of the entire body. The axial skeleton contains bones that lie on or near this midline. It consists of the skull ■, the vertebrae, the ribs, the sternum, and the hyoid bone. If the ear ossicles ■ are included, the axial skeleton contains 80 bones.

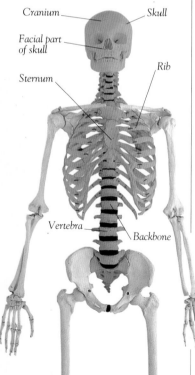

Axial skeleton
The axial skeleton (shown in red) runs down the midline of the body.

Backbone

A strong, flexible chain of bones that runs down the middle of the body

The backbone is also known as the **spine** or **vertebral column**. It consists of vertebrae that meet at joints ■. Each joint allows a small amount of movement, but together they make the backbone very flexible. Seen from the side, the backbone has four **vertebral curves**. These help to strengthen the backbone, balance the body, and absorb jolts during movement.

Vertebra

One of the bones that make up the backbone

A vertebra (plural **vertebrae**) consists of a short, pillarlike bone, called the **centrum** or **vertebral body**, attached to a ring-shaped arch, called the **vertebral arch** or **neural arch**. The centrum helps to bear the body's weight. The arch protects the spinal cord ■, which runs through a space called the **vertebral foramen**. The arch has several bony projections, called **spinous processes**. These form joints with other vertebrae, and anchor muscles ■. The backbone has 24 separate vertebrae, and nine that are partly or fully fused together.

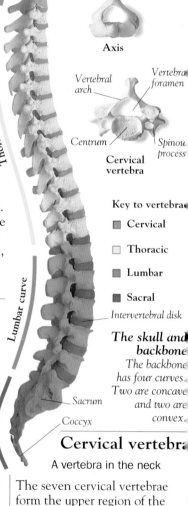

Atlas

Axis

Vertebral arch

Centrum

Vertebra foramen

Spinou process

Cervical vertebra

Key to vertebrae

■ Cervical

□ Thoracic

■ Lumbar

■ Sacral

Intervertebral disk

Sacrum

Coccyx

The skull and backbone
The backbone has four curves. Two are concave and two are convex.

Cervical vertebra

A vertebra in the neck

The seven cervical vertebrae form the upper region of the backbone. From the side, this area has an inward (concave) curve. The neck vertebrae are small and light compared to most other vertebrae. The first and second cervical vertebrae, the **atlas** and **axis**, are shaped to allow the skull to rotate and move up and down.

Thoracic vertebra

A vertebra behind the ribcage

The 12 thoracic vertebrae make up the central part of the backbone, which has an outward (convex) curve. These vertebrae form joints with the ribs.

Lumbar vertebra

A vertebra at the base of the back

The five lumbar vertebrae make up the lower, concave part of the backbone. This curve forms a hollow called the **small of the back**. Lumbar vertebrae are the largest and bear the heaviest load.

Sacrum

A triangular bone at the base of the backbone

The sacrum is made of five **sacral vertebrae**. These are separate at birth, but by 20 years of age they fuse to form a single solid bone. The sacrum has a convex curve, and provides a strong anchorage for the pelvic girdle ■.

Coccyx

A collection of small bones right at the end of the vertebral column

The coccyx is at the base of the backbone, and normally consists of four small **coccygeal vertebra**. These fuse during early adulthood.

Intervertebral disk

A pad of cartilage between neighboring vertebrae

Intervertebral disks have a jellylike center covered by fibrous cartilage ■. They allow vertebrae to move and cushion them against sudden jolts by absorbing shocks.

Slipped disk

A condition in which a disk protrudes from its normal position

If the backbone is put under a lot of pressure, a disk can be pushed out of place, or **prolapse**. Its soft center is forced through the surrounding cartilage, and may press on a nearby spinal nerve ■, preventing the nerve from working normally. Most "slipped" disks occur in the base of the back, where they can affect the sciatic nerve ■.

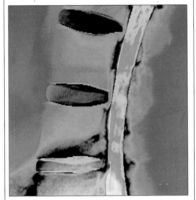

A slipped disk
This false color X-ray shows three disks (dark green) and vertebrae (light green). The disk at the bottom has prolapsed, and its soft center has been forced out.

Hyoid bone

A bone at the top of the throat

This U-shaped bone anchors muscles in the tongue and in the neck. It is not directly attached to the skeleton, and is held in place by muscles and ligaments ■.

Sternum

The bone at the front of the chest

The sternum, or **breastbone**, is a flat bone that is connected to the ribs by strips of cartilage. It has three parts. The **manubrium** is at the top and the **sternal body** is in the middle. The **xiphoid process**, at the lower end, does not have any ribs attached to it.

Rib cage

A bony frame that allows breathing and protects internal organs

The rib cage is made of 12 pairs of flat, curved bones called **ribs**. The ribs are connected to each other by intercostal muscles ■. They move up and down during breathing to make the lungs change shape. The rear end of each rib is attached to a thoracic vertebra. The front ends of the upper seven pairs, the **true ribs**, are attached to the sternum by flexible strips of **costal cartilage**. The next three pairs, called **false ribs**, are each connected to the rib above, and the lowest two pairs, called the **floating ribs**, are attached only to the backbone.

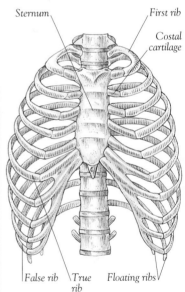

Sternum   *First rib*
Costal cartilage
False rib   *True rib*   *Floating ribs*

The rib cage
The rib cage consists of 12 pairs of ribs (yellow), most of which are connected to the sternum by costal cartilage (blue).

See also

Axis 15 • Ear ossicle 73
Fibrous cartilage 35 • Intercostal muscle 53
Joint 44 • Ligament 44 • Muscle 48
Pelvic girdle 41 • Sciatic nerve 61
Skull 38 • Spinal cord 62
Spinal nerve 61

Skull

The skull is a collection of 22 different bones in the head. It supports and protects the brain, and houses the body's most important sense organs. In adults, all but one of the skull's bones, the lower jaw, are permanently locked together and cannot move. This gives the skull its extraordinary strength.

Skull

A bony case that protects the brain and sense organs

The skull consists of a bony chamber that protects the brain ■, and a series of smaller bones that protect the sense organs ■ and form the mouth ■. It is made up of 22 bones in total – eight paired bones, and six unpaired bones. The bones are divided into two sets – the cranial bones and the facial bones. The skull also houses the bones of the inner ear ■.

Cranium

The part of the skull that surrounds the brain

The cranium is an extremely strong case made up by eight interlocking bones called **cranial bones**. The **frontal bone** forms the forehead, while the two **parietal bones** form the sides and roof of the head. Two **temporal bones** make up the part of the cranium around the ears, and the **occipital bone** forms the rear and base of the skull. The **ethmoid bone** forms part of the nasal cavity, and the **sphenoid bone** makes up part of the side of the cranium. The cranium surrounds and protects the brain. At the base of the cranium is an oval hole called the **foramen magnum**. The upper part of the spinal cord ■ passes through this hole and connects with the brain.

Side view of the skull
The photograph below shows the surface features and some of the different bones of the human skull. The skull's exterior is extremely strong, but some of its internal partitions, particularly behind the nose, are more delicate.

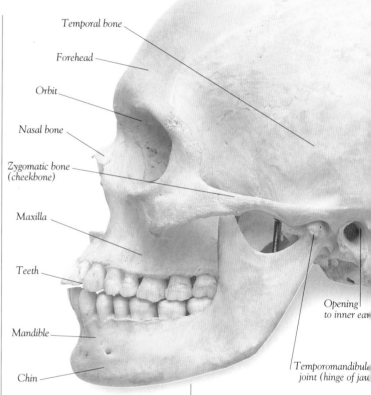

Coronal suture

Temporal bone

Forehead

Orbit

Nasal bone

Zygomatic bone (cheekbone)

Maxilla

Teeth

Mandible

Chin

Opening to inner ear

Temporomandibular joint (hinge of jaw)

Suture

An immovable joint between bones in the skull

A suture is a special kind of joint ■ found between bones in the skull. In a suture, the bones are locked together like pieces in a jigsaw puzzle. The suture prevents the bones from moving. There are four prominent sutures – the **coronal suture** across the top of the skull, the **sagittal suture**, the **lambdoid suture**, and the **squamous suture**.

Facial bone

A bone that forms part of the face

The skull contains 14 facial bones. Two of these, the **vomer** and the **mandible**, are single, while the rest are paired. The paired bones are the **nasal bones**, the **maxillae**, the **cheekbones** or **zygomatic bones**, the **lacrimal bones**, the **palatine bones**, and the **inferior nasal conchae**. Only the mandible is free to move. All the other facial bones are tightly locked together by sutures.

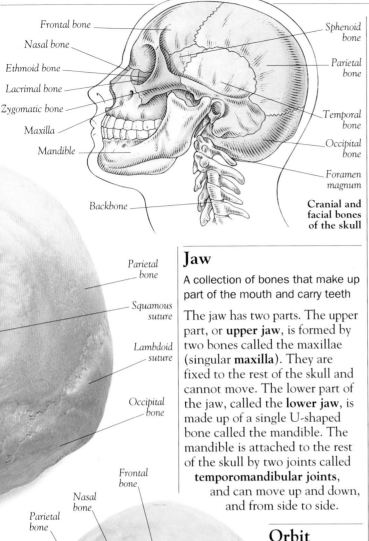

Frontal bone
Nasal bone
Ethmoid bone
Lacrimal bone
Zygomatic bone
Maxilla
Mandible
Backbone

Sphenoid bone
Parietal bone
Temporal bone
Occipital bone
Foramen magnum

Cranial and facial bones of the skull

Parietal bone
Squamous suture
Lambdoid suture
Occipital bone

Jaw

A collection of bones that make up part of the mouth and carry teeth

The jaw has two parts. The upper part, or **upper jaw**, is formed by two bones called the maxillae (singular **maxilla**). They are fixed to the rest of the skull and cannot move. The lower part of the jaw, called the **lower jaw**, is made up of a single U-shaped bone called the mandible. The mandible is attached to the rest of the skull by two joints called **temporomandibular joints**, and can move up and down, and from side to side.

Sinus

An air-filled space in the skull

Sinuses are found in many bones surrounding the nasal cavity. They help to make the skull lighter. They are lined with mucous membranes ■. **Sinusitis**, an inflammation ■ of the membranes that causes headaches and a "stuffy" head, can result from an infection ■ or an allergy ■.

Fontanelle

A gap between bones in the skull

Initially, a baby's skeleton is made of cartilage ■ and soft tissue. It slowly turns to bone in a process called ossification ■. At birth, the cranium is still not fully ossified, and it contains flexible gaps called fontanelles. These areas enable the skull to change shape as it squeezes through the mother's pelvis ■ during birth. They also allow rapid growth of the brain during infancy. The largest fontanelles are at the front and back of the skull and are called the **anterior fontanelle**, and **posterior fontanelle** respectively.

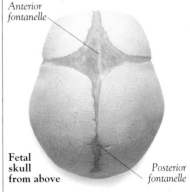

Anterior fontanelle

Fetal skull from above

Posterior fontanelle

Parietal bone
Nasal bone
Parietal bone
Temporal bone
Lacrimal bone
Sphenoid bone
Orbit
Nasal septum
Inferior nasal concha
Vomer
Maxilla
Frontal bone
Teeth
Mandible

Front view of the skull

Orbit

An eye socket

Each orbit, or **eye socket**, is formed by parts of seven different interlocking bones. The orbits protect the eyes and also form a framework for the eye muscles to pull against. They are cylindrical when seen from the front, but are actually cone-shaped.

See also

Allergy 100 • Brain 64
Cartilage 35 • Infection 92
Inflammation 94 • Inner ear 73 • Joint 44
Mouth 120 • Mucous membrane 19
Ossification 35 • Pelvis 41
Sense organ 68 • Spinal cord 62

Appendicular skeleton

The appendicular skeleton contains all the bones of the limbs, together with the bony girdles that anchor them to the rest of the body. Some of these bones are held together tightly and have limited movement. Others are free to move in many different directions.

Appendicular skeleton

The bones of the girdles and limbs

The appendicular skeleton is so called because it includes the body's **appendages**, or arms and legs. It also includes the bones that attach the arms and legs to the rest of the skeleton ▪. The appendicular skeleton contains 126 bones; over 80 percent of these are in the hands ▪ and feet ▪.

Girdle

A ring of bones that anchors the limbs to the skeleton

The body has two girdles, the pectoral girdle and the pelvic girdle. Each one anchors limbs to the rest of the skeleton, and allows them to move. The pectoral girdle is very mobile, but relatively weak. The pelvic girdle is stronger, but almost rigid.

Pectoral girdle

The girdle that anchors the arms to the skeleton

The pectoral girdle is also known as the **shoulder girdle**. Each side contains two bones – a **collarbone**, or **clavicle**, and a **shoulderblade**, or **scapula**. The clavicle is long and narrow. One end forms a joint with the sternum ▪, and the other with the scapula. The scapula is a flat, triangular bone with a hollow in one corner. The head of the humerus fits into the hollow, forming the shoulder ▪ joint.

The skeleton
The girdles and limbs that make up the appendicular skeleton enable the body to move. Despite their difference in size, the upper and lower limbs contain the same number of bones. In the skeleton shown right, the appendicular skeleton is highlighted in red.

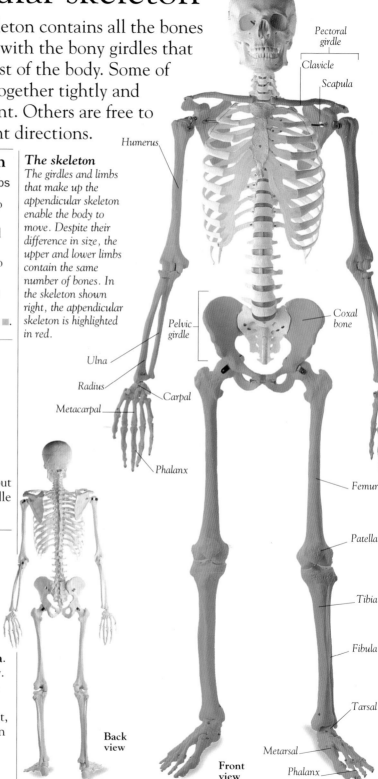

Pectoral girdle
Clavicle
Scapula
Humerus
Coxal bone
Pelvic girdle
Ulna
Radius
Carpal
Metacarpal
Phalanx
Femur
Patella
Tibia
Fibula
Tarsal
Metarsal
Phalanx

Back view
Front view

Humerus

The bone of the upper arm

The humerus is the longest bone in the arm. The end nearest the body has a rounded head that fits into the hollow of the scapula. The end furthest from the body has two joints – one with the radius, and one with the ulna. Together, these form the elbow ▪.

Ulna

The inner bone of the forearm

The ulna runs from the elbow to the wrist ▪, joining it on the little-finger side. The part of the ulna nearest the body has a hook-like tip that hinges around the humerus. The back of this hook forms the elbow's sharp point.

Radius

The outer bone of the forearm

The radius runs from the elbow to the wrist, joining it on the thumb side. The end nearest the body has a flat, disklike face. This face rotates, allowing the radius to cross over the ulna. The two bones cross when you stretch out your arms with your palms facing down, and uncross when your palms face up.

Pelvic girdle

The girdle that anchors the legs and supports the abdomen

The pelvic girdle is also called the **hip girdle**. It is made up of two **coxal bones**, or **hip bones**. These bones meet at a joint at the front of the body and are firmly connected to the sacrum ▪ at the back. Each coxal bone is actually three bones that have fused together. The **ilium** is the largest bone, and forms the hip. The smaller **pubis** and **ischium** form a ring. Together, all three bones form a rounded socket that holds the head of the femur.

Pelvis

The pelvic girdle and its neighboring bones

The pelvis is made up by the pelvic girdle, the sacrum, and the coccyx ▪. It forms a bowl that supports the abdominal organs, and contains a central space called the **pelvic inlet**. In females, a baby passes through this inlet during birth ▪.

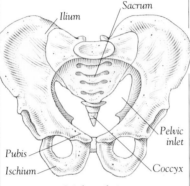

Ilium · *Sacrum* · *Pelvic inlet* · *Pubis* · *Ischium* · *Coccyx*

Male pelvis

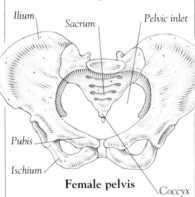

Ilium · *Sacrum* · *Pelvic inlet* · *Pubis* · *Ischium* · *Coccyx*

Female pelvis

Adult pelvis
A woman's pelvis is shallower and broader than a man's and has a wider pelvic inlet. During birth, a baby's head can just squeeze through this space.

Femur

The bone of the upper leg

The femur, or **thighbone**, is the largest bone in the body. The end nearest the body has a rounded head that fits into the pelvis. The end furthest from the body has a wide, grooved surface that forms part of the knee ▪.

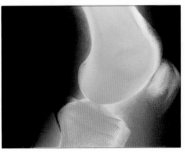

X-ray of patella
In this false color X-ray of the knee joint between the femur and the tibia, the patella is clearly visible, center right.

Patella

A bony cover that protects the knee

The patella, or **kneecap**, is a small disc-like bone that lies over the knee's surface, and protects it against damage. It forms inside a tendon ▪, which holds it in place.

Tibia

The large bone of the lower leg

The tibia, or **shinbone**, is the largest bone below the knee, and bears most of the weight in the lower leg. It runs from the knee to the ankle, and has wide ends that form part of these joints. The central part of the tibia is almost triangular, and has a sharp front edge – the **shin**.

Fibula

The small bone of the lower leg

The fibula is much smaller than the tibia, and carries very little of the body's weight. Its upper end connects with the tibia, just below the knee, and its lower end forms part of the ankle. The fibula helps to swivel the foot.

See also

Birth 140 • Coccyx 37 • Elbow 45
Foot 43 • Hand 42 • Knee 45
Sacrum 37 • Shoulder 45 • Skeleton 34
Sternum 37 • Tendon 49 • Wrist 45

Hands

Your hands are the most flexible parts of your body. They are strong enough to hold heavy weights, but delicate enough to carry out precise tasks, such as turning the pages of this book.

Hand

A part of the arm used for gripping

The hand has 27 bones. These are divided into 3 groups – the carpals in the wrist ■, the metacarpals in the palm, and the phalanges in the fingers and thumb. Over 30 muscles ■ let the hand move in different ways.

Carpal

A bone in the wrist

The eight small carpal bones are tightly held together by ligaments ■. They are arranged in two rows, and the name of each bone describes its shape. The row nearest the body has the **scaphoid** (boat-shaped), the **lunate** (crescent-shaped), the **triquetrum** (three-edged) and the **pisiform** (pea-shaped). The next row has the **trapezium** and **trapezoid** (both four-sided), the **capitate** (round-headed), and the **hamate** (hook-shaped).

Sesamoid bone

A bone that develops inside a tendon

Sesamoid means "shaped like a sesame seed." It describes any small bone that forms inside a tendon ■ as a result of prolonged pressure. Many people have sesamoid bones in their hands, in addition to the other bones.

Skeleton of the hand

This false colour X-ray shows the 27 bones in the hand. The hand may also contain extra sesamoid bones.

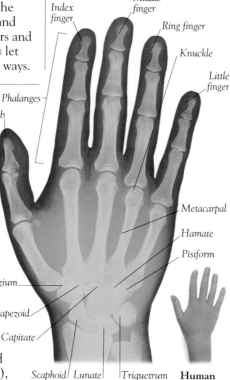

Index finger
Middle finger
Ring finger
Knuckle
Little finger
Phalanges
Thumb
Metacarpal
Hamate
Pisiform
Trapezium
Trapezoid
Capitate
Scaphoid | Lunate | Triquetrum
Human hand

Metacarpal

A bone in the palm of the hand

Five long, straight metacarpals make up the palm's framework. Each metacarpal has a joint ■ at one end with a carpal, and at the other end with a phalanx. The metacarpal in the thumb is more mobile than the others. It can swing across the palm, allowing the thumb to press against, or oppose ■, the fingers.

Phalanx of the hand

A bone in a finger or thumb

A phalanx (plural **phalanges**) is a slender bone with one or two joints. Each **finger**, or **digit**, contains three phalanges, but the **thumb**, or **first digit**, contains just two. Phalanges are also found in the foot.

Knuckle

A joint in a finger or thumb

The hand has a total of 14 knuckles – three in each finger, and two in the thumb. This large number of joints makes the hand very flexible.

Using your hands
The actions shown below demonstrate the versatility of the human hand.

Outdoor activities, such as rock climbing in which the hands partly support the body, require strength.

Everyday tasks, such as chopping vegetables, often involve skill and precision.

Painting often requires delicate movements of the hand.

See also

Joint 44 • Ligament 44 • Muscle 48
Oppose 55 • Tendon 49 • Wrist 45

Feet

The bones of your feet are among the hardest working in your whole body. They have to bear your body's weight, often for hours at a time, and also help to keep your body balanced.

Foot

A part of the leg that is used for support, balance, and movement

The foot contains 26 different bones. As with the hand, these bones fall into three groups. The tarsals are in the ankle ■, the metatarsals are in the body of the foot, and the phalanges are in the toes. Strong ligaments ■ hold the bones together so that they can bear the body's weight.

Tarsal

A bone in the ankle

The foot contains 7 tarsal bones. One of these, the talus, forms a joint ■ with the 2 bones, the tibia and the fibula, in the lower leg. The largest tarsal, the calcaneus, makes up the heel. The other 5 tarsals are smaller, and form the arch of the foot. These bones are the navicular (boat-shaped), the cuboid (cube-shaped), and the first, second, and third cuneiforms (wedge-shaped).

Metatarsal

A bone between the ankle and the toes

The five metatarsals are the longest bones in the foot. They help the foot to act as a lever during walking and running. The first metatarsal, on the inside of the foot, is thicker and stronger than the other four. It carries a large amount of the body's weight.

Phalanx of the foot

A bone in a toe

Each toe, or digit, contains bones called phalanges (the plural of phalanx). The big toe has two phalanges, but all the other toes have three. Phalanges are also found in the hand.

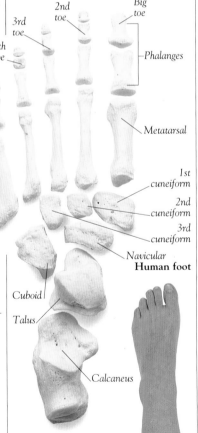

2nd toe
Big toe
3rd toe
4th toe
Little toe
Phalanges
Metatarsal
1st cuneiform
2nd cuneiform
3rd cuneiform
Navicular
Human foot
Cuboid
Talus
Calcaneus

Skeleton of the foot
The 26 bones in the foot are arranged in a similar way to the bones in the hand. However, the foot is less flexible.

Arch

A flexible curve in the foot

As the foot grows, it develops an arched shape, so that part of the sole remains off the ground. This shape spreads the weight of the body, and helps to absorb the impact with the ground during walking and running. The arch is held up by tendons ■ and ligaments.

Footprints
These footprints show the part of the foot that actually touches the ground. The curve on the inside of each footprint shows the position of the foot's arch.

Flat foot

A foot with a reduced arch

In a flat foot, the tendons and ligaments are weakened. As a result, the arch may be low or completely absent, and all of the sole may be in contact with the ground. Flat feet are quite common, and can often be corrected by special exercises.

Bunion

A swelling at the base of the big toe

A bunion is a common painful condition that affects the big toe. The joint at the base of the toe becomes thickened and bent, and the surrounding skin may be inflamed. Bunions are usually caused by wearing tight-fitting shoes but may also be inherited ■.

See also
Ankle 45 • Inherited characteristic 134
Joint 44 • Ligament 44 • Tendon 49

Joints

Almost every form of movement, from chewing food to walking or running, involves joints. Joints allow one bone to move against another, and they give the skeleton its flexibility. There are many kinds of joints in the body, each of which has a different range of movement.

Joint

A part of the skeleton where bones meet

Joints are also known as **articulations**. They are parts of the skeleton where two or more bones come into close contact. In a **fixed joint**, such as the sutures ▦ in the skull, the bones are locked together and cannot move. In a **partially movable joint**, such as the joint between neighboring vertebrae ▦, the bones can move only slightly. In a **movable joint**, the bones can move freely. People who are **double-jointed** have the same number of joints as everyone else, but their joints have a wider range of movement. In all joints, the bones are separated by layers of tissue, usually cartilage ▦.

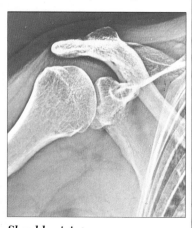

Shoulder joint
This false color X-ray shows the three bones that meet to form the shoulder joint. The gap between them is filled by cartilage and synovial fluid.

Synovial joint

A movable joint with a fluid-filled space

Synovial joints allow the body to move easily. They include the shoulder, elbow, hip, and knee, and all the joints in the fingers and toes. The bones in a synovial joint are tipped with a layer of smooth hyaline cartilage ▦, and the cartilage is separated by a space containing an oily yellow fluid called **synovial fluid**. Synovial fluid lubricates the cartilage and allows bones to slide over each other more easily. The entire joint is surrounded by a tough covering called a **capsule**. The capsule is lined by a **synovial membrane**, which produces the synovial fluid.

Meniscus

A pad of cartilage inside a synovial joint

A meniscus (plural **menisci**) is a crescent-shaped piece of cartilage that is found in some synovial joints. The knees, the wrists, and the hinge of the jaw all contain menisci. They are attached to a joint's capsule, and they help to reduce friction between the bones. Menisci in the knee can become damaged during strenuous activity, and sometimes have to be removed to prevent the joint from locking.

Ligament

A strip of fibrous connective tissue that holds bones together

Ligaments link neighboring bones where they meet at joints. Some ligaments hold bones together so tightly that they can hardly move. Others are looser and allow the bones to change position. Ligaments are extremely strong. Although they sometimes tear, it takes a very powerful jolt to break them.

Gliding joint

A joint that allows bones to slide from side to side

Gliding joints are also known as **plane joints**. In this kind of joint, the bone surfaces are practically flat and they move by sliding over each other. Gliding joints are found between the carpals ▦ in the hand and between the tarsals ▦ in the foot.

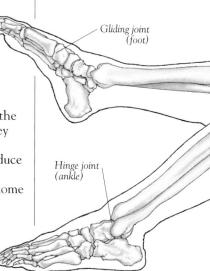

Gliding joint (foot)

Hinge joint (ankle)

Hinge joint

A joint that allows movement in one plane

In a hinge joint, a cylindrical surface fits inside a curved recess. The bones can move up and down, but not from side to side. The **knee** is a hinge joint, made up by the femur ■, fibula ■, and tibia ■. The talus ■, tibia, and fibula bones form the hinge joint of the **ankle**, and the joints between the phalanges ■ in the hands and feet are also hinge joints. The **elbow** is a rotating hinge joint formed by the humerus ■, ulna ■, and radius bones ■.

Saddle joint

A joint that allows movement in two planes

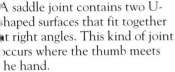

A saddle joint contains two U-shaped surfaces that fit together at right angles. This kind of joint occurs where the thumb meets the hand.

Ellipsoidal joint

An oval joint that allows movement in two directions

An ellipsoidal joint is also called a **condyloid joint**. It has an oval head that fits into an oval cup. The bones can move backward, forward, or from side to side. The joint between the radius and the carpals in the **wrist** is an ellipsoidal joint.

Ball-and-socket joint

A joint that allows movement in many directions

A ball-and-socket joint contains a bony ball that fits inside a cup-shaped socket. It is the most flexible kind of joint in the body, and is found in the **hip** joint, between the pelvic girdle ■ and the femur, and in the **shoulder** joint, between the pectoral girdle ■ and the humerus.

Pivot joint

A joint that allows a bone to rotate

In a pivot joint, one bone swivels inside a space formed by another. The joint in the neck between the axis ■ and atlas ■ vertebrae is an example of a pivot joint.

Joints of the body
Joints that have a wide range of movement tend to be weaker than those that are less flexible.

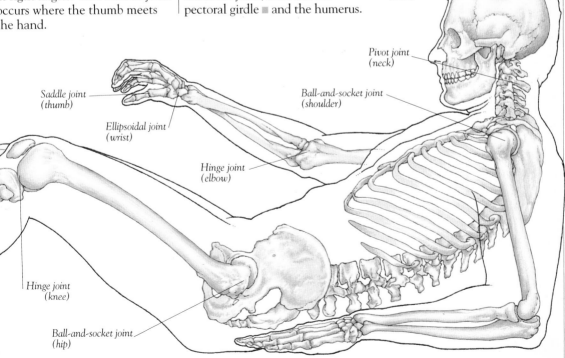

Fixed joint (skull)

Pivot joint (neck)

Saddle joint (thumb)

Ellipsoidal joint (wrist)

Ball-and-socket joint (shoulder)

Hinge joint (elbow)

Hinge joint (knee)

Ball-and-socket joint (hip)

Joints & movement

Bending your back is an example of a movement called flexion, while straightening your leg is an example of extension. These are just two of the movements that joints, together with bones and muscles, allow the body to perform.

Flexion

A movement that reduces the angle of a joint

When you bend, or **flex**, a joint, you make its bones come closer together. Flexing your elbow brings your upper arm and forearm closer together. Flexing your fingers brings them closer to the rest of your hand. The muscles that flex joints are known as **flexors**. The flexor digitorum brevis ■, for example, is a muscle that flexes four of your toes.

Extension

A movement that increases the angle of a joint

During extension, a joint opens out so that its bones move further apart. For example, when you kick, you straighten, or **extend**, your knee joint, and when you point at something, you extend your finger joints to straighten them. Muscles that extend joints are called **extensors**.

See also

Adductor longus 56 • Cartilage 35
Depressor labii inferioris 50
Femur 41 • Flexor digitorum brevis 57
Immune system 98
Intercostal muscle 53 • Ligament 44
Pelvis 41 • Platysma 51 • Tibia 41

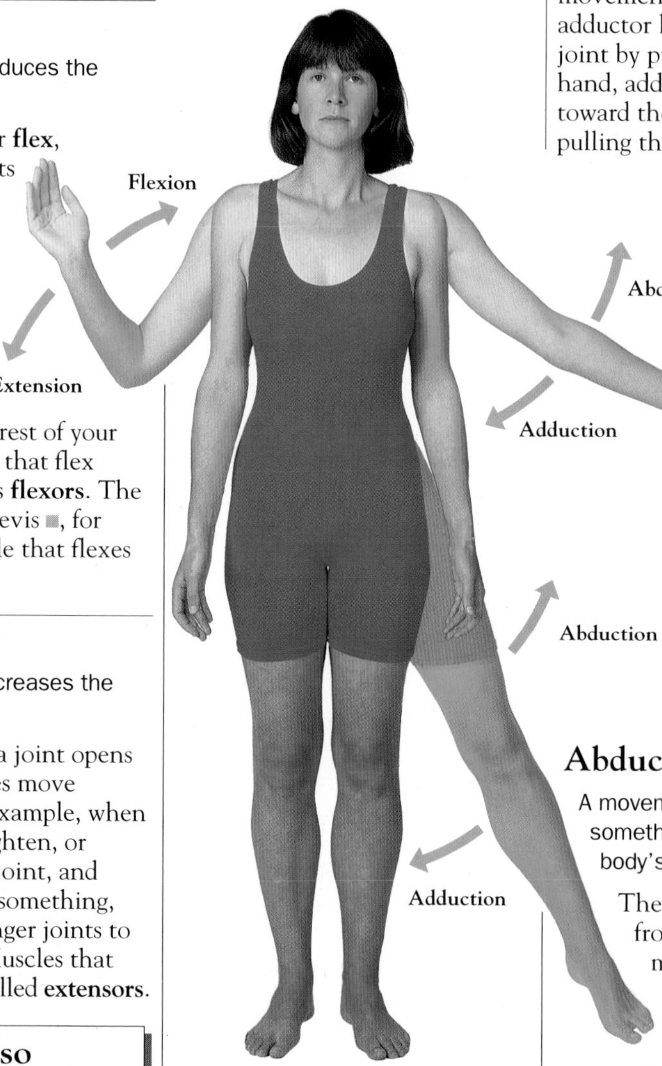

Flexion

Extension

Abduction

Adduction

Adduction

Abduction

Adduction

Body movements
In this picture, the arrows show different types of joint movement. The woman's right arm flexes and extends at the elbow. Her left arm and leg demonstrate movement away from the body's midline (abduction), and movement toward the body's midline (adduction).

Adduction

A movement that takes something toward the body's midline

Adduction comes from the Latin words meaning "to lead toward." It usually refers to any movement toward the body's axis ■, or midline. An **adductor** is a muscle that produces this kind of movement. For example, the adductor longus ■ moves the hip joint by pulling in the leg. In the hand, adduction is a movement toward the midline of the hand, pulling the fingers together.

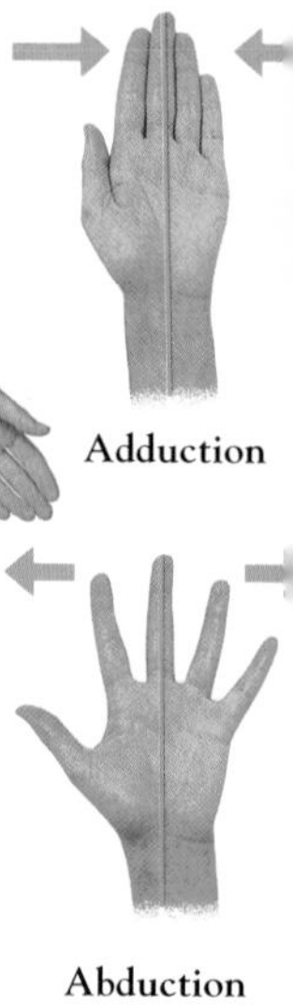

Abduction

Adduction

Abduction

Abduction

Abduction

A movement that takes something away from the body's midline

The term abduction comes from the Latin words meaning "to lead away." During abduction, the movement of a joint makes something swing away from the body's midline. Fingers and toes can also be abducted, but they move away from the midline of the hand or foot, rather than from the midline of the body. An **abductor** is any muscle that produces a movement away from the midline.

Depression

A movement that pulls something downward

Used to describe movement, depression refers to a downward motion. A muscle that produces this kind of movement is called a depressor. For example, the platysma ▪ helps to pull down, or depress, the lower jaw. The depressor labii inferioris ▪ pulls down the lower lip.

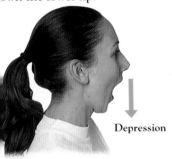

Depression

Lowering the jaw
This picture demonstrates the lowering, or depression, of the jaw bone, or mandible.

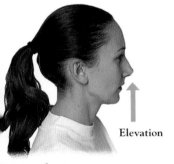

Elevation

Raising the jaw
This picture demonstrates the raising, or elevation, of the jaw bone, or mandible.

Elevation

A movement that lifts something upward

During elevation, part of the body is raised from its normal position. For example, during breathing, the intercostal muscles ▪ raise, or elevate, the ribs, so that air is sucked into the lungs. A muscle that lifts something in this way is known as a **levator**.

Dislocation

A movement of a bone from its normal position in a joint

When a joint is dislocated, the bones are wrenched so hard that they are pulled or pushed out of place. The joint cannot work, and treatment is usually needed to put the bones back into position. Dislocation is common in the shoulder, and also in the joints of the fingers and thumbs.

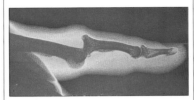

Dislocated finger
This false color X-ray shows the dislocation of two bones, or phalanges, in a finger. Ligaments around a joint are often torn during dislocation, which is why this injury can be so painful.

Sprain

A tearing of the ligaments in a joint

Joints are built to withstand tough treatment. However, if a joint is forced beyond its normal limits of movement, such as a sudden pull, an injury can occur. In a sprained joint, one or more ligaments ▪ become torn. If the sprain is severe, the joint may have to be kept still for several weeks until the ligaments repair themselves.

Rheumatism

Any condition producing pain or stiffness in joints and muscles

The term rheumatism refers to disorders that cause pain and stiffness in joints and surrounding muscles. It includes minor aches and pains that cause no lasting harm, as well as more serious forms of rheumatoid arthritis that make joints stiff, swollen, and painful.

Arthritis

A disease that affects the joints

Arthritis is not a single disease, but a range of diseases and disorders that make joints less mobile. In **osteoarthritis**, the cartilage ▪ in joints begins to break down, which makes the joints stiff and difficult to move. This kind of arthritis is common in people over 60 years old. In **rheumatoid arthritis**, the joints become swollen and painful. If the disease is severe, the joints may become deformed. Rheumatoid arthritis is caused by an immune system ▪ disorder, and usually affects people from early middle age onward.

Artificial joint

A manufactured replacement for one of the body's joints

If a joint becomes badly diseased, it may be replaced by an artificial joint. In a **hip replacement**, the top of the femur ▪ is replaced by a metal ball. This rotates within a plastic socket inserted into the pelvis ▪. In a **knee replacement**, the femur and tibia ▪ are joined by an artificial hinge, or capped by plastic covers that replace worn-out cartilage.

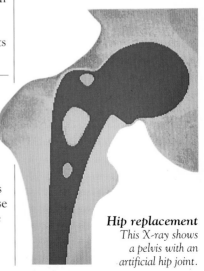

Hip replacement
This X-ray shows a pelvis with an artificial hip joint.

Muscles

Whether you are running a marathon or sound asleep in bed, your muscles are at work. Muscles convert chemical energy into movement and heat. Some muscles work only when you want them to, but others work continuously to keep your body functioning.

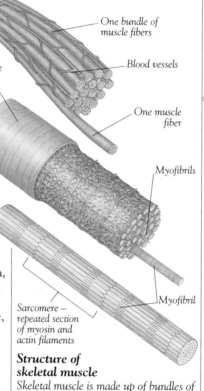

Whole muscle

Bundles of muscle fibers

One bundle of muscle fibers

One muscle fiber

Blood vessels

One muscle fiber

Myofibrils

Myofibril

Sarcomere – repeated section of myosin and actin filaments

Muscle

A tissue that contracts to produce movement or tension

Muscles are made up by long cells called **muscle fibers**, or **myofibers**. A single fiber contains thousands of tiny strands called **myofibrils**. Each myofibril contains filaments of two proteins, called **actin** and **myosin**. When the muscle fibers receive a signal – usually from a nerve ■ – the fibers shorten, making the muscle contract. The force of the contraction moves a part of the body, holds it in place, or changes its shape. There are three types of muscle – cardiac, skeletal, and smooth. In cardiac and skeletal muscle, the actin and myosin filaments do not run the full length of the myofibril but are arranged into repeated patterns or **sarcomeres**. Smooth muscle is less regular and has no sarcomeres.

Structure of skeletal muscle
Skeletal muscle is made up of bundles of muscle fibers. Each fiber contains many myofibrils, which in turn contain thick myosin filaments and thin actin filaments.

Cardiac muscle

The muscle of the heart

Cardiac muscle, or myocardium, is found in the heart. It has oblong branched fibers, which are striated – they look striped when magnified. Cardiac muscle contracts and relaxes by itself 100,000 times a day and never gets tired. It has its own built-in rhythm, but this can be altered by nerves and some hormones ■.

See also

Aerobic respiration 104
Alimentary canal 116
Anaerobic respiration 104 • ATP 104
Autonomic nervous system 60
Connective tissue 19 • Glucose 22
Homeostasis 76 • Hormone 78
Lactic acid 104 • Nerve 58
Peristalsis 116 • Vitamin 108

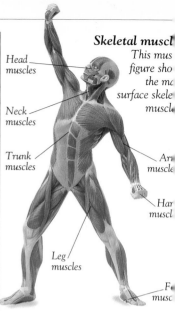

Skeletal muscl
This mus figure sho the m surface skele muscl

Head muscles

Neck muscles

Trunk muscles

Leg muscles

Ar muscle

Har muscl

F musc

Skeletal muscle

Muscle that is attached to the skeleton

Skeletal muscles are also called **voluntary muscles**, because you can consciously decide to make them contract or relax. They ar connected to the skeleton, and enable your body to move. The fibers of skeletal muscle are long and striated. The human body has over 600 individually name skeletal muscles. They make up about 40 percent of the body's weight and generate a lot of heat. Skeletal muscles can pull, but cannot push. To make thing move, they are often arranged ir opposing pairs or groups, called **antagonistic pairs**. The muscles in each pair work against each other, so when one contracts, the other relaxes.

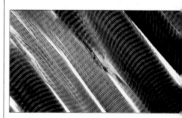

Skeletal muscle
The fibers of skeletal muscle lie paralle to each other and are striped, or striate

Smooth muscle

Muscle that is found in the wall of hollow organs inside the body

Smooth muscle is also known as **involuntary muscle**, because you cannot consciously decide to make it contract. Instead, it is triggered by the autonomic nervous system ■, and by hormones. Its fibers have tapering ends, and they are **unstriated**, which means that they do not look striped under a microscope. Smooth muscle forms a double layer around hollow parts of the body, such as the alimentary canal (digestive tract) ■ and blood vessels. The fibers in the two layers run in different directions, so they have opposite effects when they contract. Smooth muscle moves food through your digestive system by peristalsis ■, and also carries out many processes of homeostasis ■.

Smooth muscle
The cells of smooth muscle are spindle-shaped and are packed closely together.

Tendon

A tough cord that connects skeletal muscles to bones

Tendons link muscles to bones and also to other muscles. They are made of strong connective tissue ■. When a muscle contracts, the tendon pulls against the bone, making the bone move at the joint. Some tendons are broad and flat but most are shaped like narrow cables. The tendons in the hands and feet are long and are enclosed by slippery **synovial sheaths**, which help them to slide when fingers or toes bend.

Muscle contraction

The shortening of a muscle

Muscles contract in two ways. During an **isotonic contraction**, muscles exert a steady pull and get much shorter. During an **isometric contraction**, muscles exert a strong pulling force, or **tension**, but become only slightly shorter. You would use an isotonic contraction to pick up this book, and an isometric contraction to hold it steady. Regardless of how relaxed you feel, your muscles are always slightly contracted to maintain your body's **posture**. This partial contraction produces firmness, or **muscle tone**, without which your body would collapse under the influence of gravity.

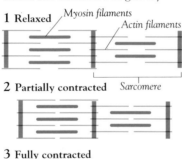

1 Relaxed Myosin filaments Actin filaments

2 Partially contracted Sarcomere

3 Fully contracted

Theory of muscle contraction
As a muscle contracts, the actin filaments slide towards the center of the sarcomere and further overlap with the myosin filaments, and sometimes with each other.

Sliding filament theory

A theory that suggests how a muscle contracts

This theory suggests that when a muscle is stimulated by a nerve, the actin and myosin filaments stay the same length but slide past each other. This shortens the myofibrils and makes the muscle contract. When the nerve signal stops, the filaments slide back, the myofibrils return to their original length and the muscle relaxes.

Albert von Szent-Györgi

Hungarian-American biochemist (1893–1986)

Albert Szent-Györgi carried out some important research into the way muscles contract. In one experiment, he prepared a mixture of the proteins actin and myosin, and then added the energy carrier ATP ■. The proteins suddenly contracted, as they do in living muscle. Szent-Györgi's research helped scientists to understand how muscles work. He also isolated vitamin C ■, and was awarded the Nobel Prize for this in 1937.

Muscle fatigue

The gradual weakening of a muscle's pull

Muscles work by combining glucose ■ with oxygen to release energy, in a process called aerobic respiration ■. However, during vigorous exercise, muscles often use up oxygen faster than the body can supply it. When this happens, they have to carry out anaerobic respiration ■, which creates a waste product called lactic acid ■. Lactic acid builds up in muscle tissue, and stops it working efficiently. Until the muscle rests and oxygen becomes available, the lactic acid cannot be broken down.

Cramps

An unwanted muscle contraction

Cramps are the sudden and often painful contractions of muscles. It can occur after exercise, and also during sleep. Probable causes include the buildup of lactic acid, and salt loss through sweating.

Face & neck muscles

The face and neck contain one of the most varied collections of muscles in the whole body. More than 30 small facial muscles express a huge range of feelings, from pleasure and surprise to anxiety and anger. Much more powerful muscles move the jaw, and keep the head upright. A selection of these muscles are described here.

Frontalis

A muscle that wrinkles the forehead

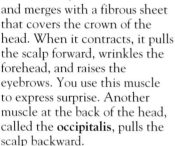

The frontalis muscle covers the whole of the forehead, and merges with a fibrous sheet that covers the crown of the head. When it contracts, it pulls the scalp forward, wrinkles the forehead, and raises the eyebrows. You use this muscle to express surprise. Another muscle at the back of the head, called the **occipitalis**, pulls the scalp backward.

Corrugator supercilii

A muscle that pulls the eyebrow down

"Corrugo" means "wrinkle", and supercilium means "eyebrow." One corrugator supercilii muscle lies above each eye, and when they both contract, they pull the eyebrows down and toward each other. You use these muscles when you frown, and also when you are exposed to very bright light. By pulling your eyebrows down, the muscles help to shade your eyes from the light.

Orbicularis oculi

A muscle that closes the eye

"Orb" means circular, while "oculi" means "of the eye." One of these muscles is anchored to each eye socket, or orbit ▦. When they contract, they make the eyelids close. They can move very quickly, enabling you to blink. Beneath each muscle is a **levator palpebrae superioris** muscle, or "lifter of the upper eyelid," which opens the eye wide.

Orbicularis oris

A muscle that purses the lips

"Oris" means "of the mouth," and this muscle forms a ring around the mouth. When it contracts, the lips tighten, close, and press against the teeth. You use the orbiculari oris when you speak and when you purse your lips. Two other muscles pull the lips apart. The two **levator labii superioris** muscles, one on either side of the nose, raise the upper lip. The two **depressor labii inferiori** muscles, one on each side of the chin, pull down the lower lip.

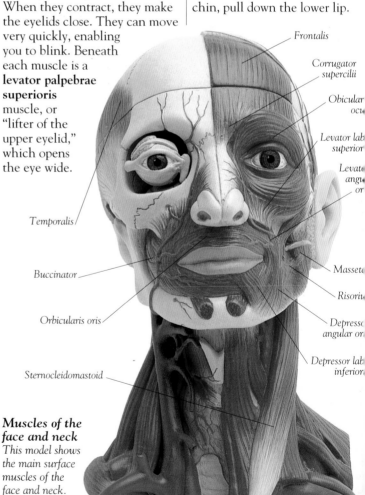

Muscles of the face and neck
This model shows the main surface muscles of the face and neck.

Frontalis

Corrugator supercilii

Obicular ocu

Levator lab superior

Levate angi or

Temporalis

Buccinator

Orbicularis oris

Sternocleidomastoid

Masset

Risoriu

Depresso angular or

Depressor lab inferior

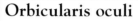

...isorius

...muscle that
...etches
...orner of
...e mouth

...he two risorius
...scles, one
... either side
... the mouth, stretch the
...outh wide when you smile or
...ugh. Two other muscles move
...e corners of the mouth. At
...her edge of the mouth, a
...vator anguli oris muscle turns
...e corner upward. On either
...le of the chin, a **depressor
...guli oris** muscle turns the
...rner downward.

...atysma

...muscle that
...rns down and
...retches the
...outh

... single, large
...atysma
...uscle runs
...wn either
...le of the chin and neck. Each
... shaped like an upside-down
...n. The wide end of the fan is
...tached to the shoulder and
...e upper chest, while the
...rrow end is attached to the
...in around the jaw and mouth.
... you turn the edges of your
...outh down as far as they will
..., you will be able to feel your
...atysma muscles tightening
... your neck.

...uccinator

...muscle that tightens the cheek

...he buccinator is the major
...uscle in each cheek. Together,
...ese muscles are used for
...owing and sucking. Each
...uccinator links the orbicularis
...ris muscle around the mouth to
...e upper and lower jaw. When
...e buccinator contracts, the
...eek is pulled tight.

Masseter

A muscle used to bite and chew

On each side of the face, a single
masseter muscle runs from the
lower jaw, or mandible ■, to the
cheekbone, or zygomatic bone ■.
Each masseter works with one of
the two **temporalis** muscles,
fixed to the temporal bones ■, to
pull the jaw upwards. Together,
these muscles exert an enormous
force. When you bite, the muscles
on both sides of the jaw pull
equally. When you chew, they
pull unequally, moving the jaw
from side to side. Small muscles
called the **pterygoids** below the
masseter help with this movement.

Sternocleidomastoid

A muscle that pushes the head
forward or makes it turn

A single sternocleidomastoid
muscle runs from the breastbone,
or sternum ■, to a point just below
the ear on each side of the neck.
Each one is also attached to the
clavicle ■, and to the temporal
bone in the skull ■. When both
muscles contract, they pull the
head forward. When only one
contracts, it makes the head turn.

Longissimus capitis

A muscle that lifts or turns the head

The name of this muscle means
"longest of the head." A single
longissimus capitis muscle on each
side of the neck reaches down the
back of the neck, from the
temporal bone in the skull, to the
lowest cervical vertebrae ■ in the
backbone ■. Without support, the
head hinges forward, as you can
see when someone "nods off to
sleep." The longissimus capitus
and several other neck muscles
keep the head upright and enable
it to turn. The head is also turned
and tilted by the **splenius capitis**
muscles, while the **splenius
cervicis** muscles twist the neck.

Rectus eye muscle

A muscle that moves the eyeball
vertically or horizontally

Six external muscles (four rectus
and two oblique muscles) move
each eyeball. The **lateral**, **medial**,
inferior, and **superior rectus**
muscles move the eyeball up and
down, or from side to side. They
are arranged at right angles to
each other, and run from the
eyeball to the back of the orbit.

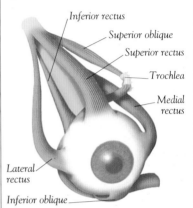

Inferior rectus
Superior oblique
Superior rectus
Trochlea
Medial rectus
Lateral rectus
Inferior oblique

Rectus and oblique eye muscles
*Six external eye muscles move the
eyeball. Hidden internal muscles change
the shape of the lens and iris.*

Oblique eye muscle

A muscle that moves the eyeball
diagonally

Each eye has two external oblique
muscles. The **inferior oblique**
runs around the underside of the
eyeball to a part of the orbit near
the nose. The **superior oblique**
runs forward from the eyeball,
and doubles back through a loop
of cartilage called the **trochlea**.
The trochlea works like a pulley,
so that the eyeball can swivel
downward and outward.

Trunk muscles

Layer upon layer of different muscles
surround the central part of your body.
Of the dozens of muscles found here, some
enable you to move or breathe, while others
prevent your body sagging or crumpling
under its own weight.

Pectoralis major

A chest muscle that pulls the arm
in toward the body

The two pectoralis major muscles
are situated on either side of the
chest. Commonly known as the
pectoral muscle, the pectoralis
major is fan-shaped and split into
two unequal parts. At its narrow
end, both parts are attached to
the humerus ▪. At its wide end,
the smaller part is attached to
the clavicle ▪, and the larger part
is attached to the sternum ▪ and
the costal cartilage on the ribs ▪.
The pectoral muscle pulls the
arm forward and in toward the
body. It also rotates the arm.

Deltoid

A shoulder muscle that moves
the arm

The powerful deltoid muscle
wraps around the shoulder, and
connects three bones – the
scapula ▪, the clavicle, and the
humerus. Most movements of
the shoulder and the upper arm
involve the deltoid. It steadies
the shoulder, and moves the arm
in many directions. Deltoid
means "triangular."

Weightlifting
*As well as using
their arm muscles,
weightlifters make
use of chest
muscles such as
the pectorals, and
shoulder muscles
such as the deltoids.*

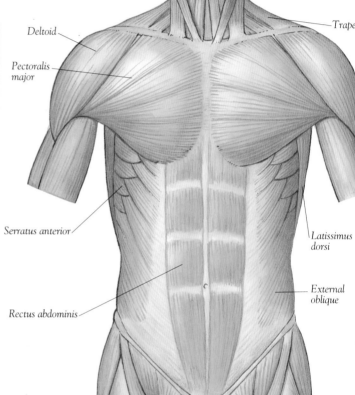

*Muscles of t
front of the tru*
*The surface mus
of the front of
trunk, as shown he
help to lift the ar
and move the b
forward and sidewa
They also protect
organs inside the abdom*

Deltoid

Pectoralis
major

Serratus anterior

Rectus abdominis

Trapez

Latissimus
dorsi

External
oblique

Serratus anterior

A chest muscle that rotates
the scapula

Each of the two serratus anterior
muscles runs from the upper ribs
at the front of the chest,
around the side, to
the scapula. The
serratus anterior
pulls the scapula
outward, which
lifts the shoulder.
Serratus means "saw-
shaped," and describes
this muscle's jagged shape.

Rectus abdominis

A muscle that tightens the abdom

The organs of the lower trunk,
abdomen ▪, are protected and
held in place by several sheet-
like muscles. These include the
two rectus abdominis muscles,
which run down the front of th
body, from the ribcage to the
front of the pelvis ▪. When the
muscles contract, the abdomen
pulled in. By working together,
the abdominal muscles allow th
trunk to move forward or
sideways, and also to twist.

External oblique

A diagonal muscle that tightens the abdomen

The two external oblique muscles run from the lowest ribs to the body's front midline ■. Here, they join to form a thin, strong fibrous sheet. These muscles help the rectus abdominis and a deeper set of oblique muscles, the **internal obliques**, to keep the abdominal organs in place.

Trapezius

A muscle in the back that pulls the head and shoulders backward

The two trapezius muscles extend from the backbone ■ and base of the skull ■, across the back and shoulders to join the scapula and the clavicle. They lift and tilt the head, and lift or steady the shoulders. Together, they make up a flat, four-sided shape called a trapezium, which gives them their name.

Trapezius

Latissimus dorsi

Muscles of the back of the trunk
The surface muscles of the back of the trunk, as shown here, help to move the arms and the head, and also to hold the back upright.

Latissimus dorsi

A back muscle that pulls the arm downward and backward

Each of these two triangular muscles has a wide end attached to the backbone and pelvic girdle ■. A narrower end runs under the shoulder to the humerus. The name latissimus dorsi means "widest of the back." This muscle supports the arm when it is raised above the head, and moves the arm down from a raised position. If you push your arm hard against your side, you will feel this muscle tighten up.

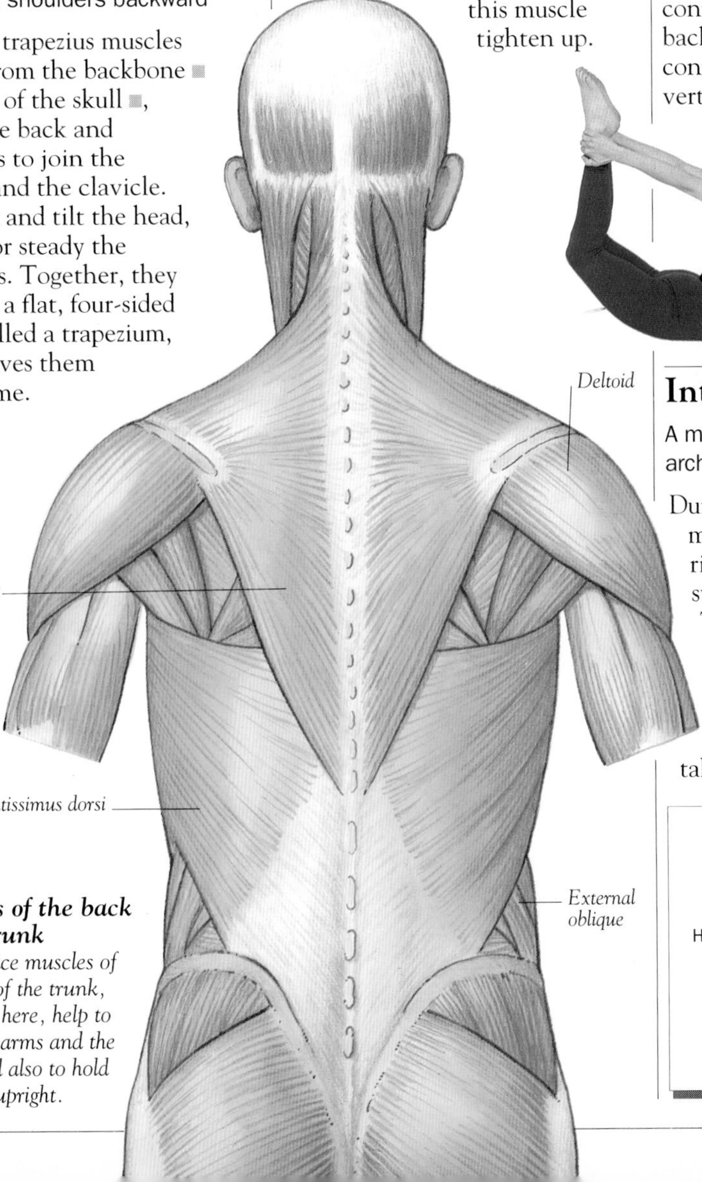

Deltoid

External oblique

Spinalis thoracis

A muscle that helps to support the back in an upright position

If you make your back as hollow as you can, you will be able to feel a set of spinalis thoracis muscles running down either side of your backbone. These muscles are attached to your vertebrae ■, and they enable you to stay upright. The spinalis thoracis muscles are connected mainly to the spinous processes ■ of the thoracic vertebrae ■ near the top of the back, and this is how they get their name. Other muscles connect different regions of the backbone. There are also muscles connecting neighboring vertebrae, enabling the backbone to twist.

Bending back
The spinalis thoracis muscles enable the back to bend, as shown here. In this curved position they stand out on either side of the backbone.

Intercostal muscle

A muscle that connects the bony arches of two neighboring ribs

During breathing, the intercostal muscles between each pair of ribs contract, making the ribs swing outward and upward. They work together with the diaphragm ■ to draw air into the lungs ■. If you exercise hard, muscles in your neck and abdomen help you to take deeper breaths.

See also

Abdomen 14 • Backbone 36
Clavicle 40 • Diaphragm 114
Humerus 41 • Lung 112 • Midline 15
Pelvic girdle 41 • Pelvis 41
Rib 37 • Scapula 40 • Skull 38
Spinous process 36 • Sternum 37
Thoracic vertebra 37 • Vertebra 36

Arm & hand muscles

The muscles of the arm and hand work together. Whether they are delicately threading a needle or smashing a tennis ball across a net, they exert exactly the right amount of pressure at the right point, and allow an amazing amount of flexibility and precision.

Brachioradialis

A muscle that raises and rotates the forearm

If you hold out your arm with the palm facing upwards, the brachioradialis forms a bulge on the outside edge of your forearm This muscle connects the radius and humerus at the ends nearest the hand. It raises the forearm and makes it rotate.

Deltoid

Biceps brachii

Brachioradialis

Flexo digitoru superficia

Flexor carpi ulnaris

Flexor carpi radialis

Palmaris longus

Triceps brachii

Brachialis

Pronator teres

Arm and hand muscle

The main surface, or superficial muscles of the arm and hand are show here. Other deeper muscles and tendon lie below these muscles

Biceps brachii

A muscle that bends the arm

The biceps connects the shoulderblade, or scapula ▪, to the radius ▪ in the forearm. When it contracts, it makes the arm bend, or flex ▪. The name biceps brachii means "two heads of the arm," and the end of the muscle nearest the body is split into two parts. Bodybuilders who want to show off their muscles often flex the biceps because it bulges visibly when it contracts.

Biceps in action
The biceps in this tennis player's arm contracts to raise his forearm and lift the tennis racket.

Brachialis

A muscle that raises the forearm

The brachialis runs from the humerus ▪ to the ulna ▪, and lies underneath the biceps. When the brachialis contracts, it raises the forearm so that the elbow fully or partly flexes.

Triceps brachii

A muscle that straightens the arm

The triceps runs down the back of the upper arm. It straightens, or extends ▪, the arm. The word "tri" shows that it has three heads – one is attached to the scapula, and the other two to the humerus. The far end of the muscle is attached by a strong tendon ▪ to the ulna, at the point of the elbow. If you make your arm as straight as possible, you can feel this tendon tighten up.

Pronator

A muscle that turns the palm of the hand downward

The **pronator teres** and **pronato quadratus** are two small muscles that run across the forearm from the ulna to the radius. When they contract, they turn the arm so that the palm faces downward, or becomes **prone**.

Supinator

A muscle that turns the palm of the hand upward

The supinator muscle is a deep muscle in the upper part of the forearm. It runs from the ulna and humerus to the radius. When it contracts, it pulls on the radius and turns the palm upward.

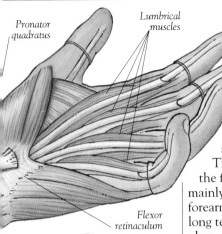

Pronator quadratus

Lumbrical muscles

Flexor retinaculum

Flexor of the hand

A muscle that bends the wrist or fingers

If you press down hard with your fingertips, you will feel flexor ■ muscles contracting on the underside of your forearm. These six muscles are attached near the elbow, and have long tendons that connect them to parts of the hand. The tendons pass under a strong ligament called the **flexor retinaculum**, which holds them to the wrist. The flexors, except for the **palmaris longus**, are named by their function, size, or position. For example, the **flexor digitorum superficialis** is a surface, or superficial, muscle, that flexes the fingers, or digits ■. Other flexors include the **flexor carpi ulnaris** and **flexor carpi radialis**.

Extensor of the hand

A muscle that straightens the hand or fingers

The nine extensor ■ muscles pull the hand back at the wrist, or straighten the fingers. They are antagonistic ■ to the flexor muscles, and they lie mainly in the upperside of the forearm. The extensors have long tendons that are held in place at the wrist by a ligament called the **extensor retinaculum**.

Palmaris brevis

A muscle that draws the skin into the palm of the hand

The palmaris brevis is a small muscle that pulls in the skin on the underside of the hand. This makes your palm form a cuplike shape. The muscle's name means "short [muscle] of the palm."

Lumbrical muscle

A narrow muscle that helps to extend the fingers

Lumbrical means "wormlike," and describes four muscles that lie in the middle of the palm. The lumbricals are attached to tendons, and help to straighten the fingers.

Interosseus muscle

A muscle that enables the fingers or toes to spread out or close up

Interosseus (plural **interossei**) means "between bone." It is the name given to eight small deep muscles that lie alongside the metacarpal ■ bones in the palm. When the interossei muscles contract, they make the fingers move sideways.

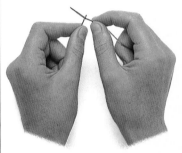

Opposing finger and thumb
Humans and apes are the only animals that can make their fingers and thumb press against, or oppose, each other, as shown here when threading a needle.

Opponens pollicis

A muscle that pulls the thumb across the palm

The thumb is the most mobile part of the hand. It is moved by four muscles in the forearm, and four in the hand itself. The opponens pollicis is one of the muscles that forms a bulge at the thumb's base. Its name means "opposer of the thumb." It moves the thumb so that it can press against, or **oppose**, the fingers. Other thumb muscles include the **flexor pollicis brevis**, which bends the thumb, and the **abductor pollicis brevis**, which moves the thumb outward.

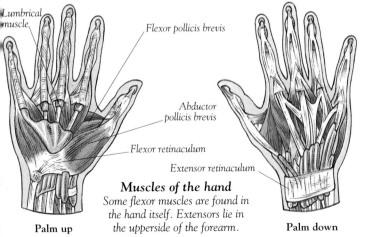

Lumbrical muscle

Flexor pollicis brevis

Abductor pollicis brevis

Flexor retinaculum

Extensor retinaculum

Muscles of the hand
Some flexor muscles are found in the hand itself. Extensors lie in the upperside of the forearm.

Palm up

Palm down

See also

Antagonistic pair 48 • Digit 42 • Extend 46
Extensor 46 • Flex 46 • Flexor 46
Humerus 41 • Metacarpal 42 • Radius 41
Scapula 40 • Tendon 49 • Ulna 41

Leg & foot muscles

Your legs are the most muscular parts of your body. They provide the power to make you move, and help to hold you steady when you stand. Foot muscles are similar to those in the hands, but the feet are stronger and less flexible.

Quadriceps femoris

A set of muscles that straightens the knee

Quadriceps means "four heads," and this muscle is actually made up of four large muscles: the **rectus femoris**, the **vastus lateralis**, the **vastus medialis**, and the **vastus intermedius**. When it contracts, the quadriceps pulls on tendons ■ that lie over the knee, and makes the knee straighten. You use this muscle to walk and run.

Stretching the leg
As this hurdler jumps, the quadriceps femoris contracts to stretch his front leg out.

Sartorius

A muscle that turns and bends the leg

The sartorius is the longest muscle in the body. It is strap-shaped, and winds across the front of the thigh, from the hip to the inner side of the tibia ■. When it contracts, it bends the leg and rotates the thigh.

Adductor longus

A muscle that pulls the leg toward the body's midline

This muscle lies on the inside of the thigh and connects the bottom of the pelvis ■ to the central part of the thighbone, or femur ■. When it contracts, it pulls the leg inwards. Three other muscles are involved in this movement: the **adductor brevis**, the **adductor magnus**, and the **pectineus**.

Extensor of the foot

A muscle that makes the foot or toes move upward

The four foot extensor ■ muscles are in the front and outer side of the lower leg. When they contract, they pull the foot or toes upward. These muscles have long tendons, which bend around the angle of the ankle where the leg meets the foot, and are held in place by the ligaments of the **extensor retinaculum**. The extensors of the foot are the tibialis anterior, the **extensor hallucis longus**, which lifts the big toe, the **extensor digitorum longus**, which lifts the other toes, and the **peroneus tertius**, which lifts the foot. On the top of the foot the only muscle is the **extensor digitorum brevis**, or "short extensor of the digits." This works with extensors in the lower leg to lift the toes.

Tibialis anterior

An extensor muscle that straightens or lifts the foot

This muscle runs alongside the tibia, and connects the upper part of the tibia with two bones on the inner edge of the foot's arch ■. When the tibialis anterior contracts, it lifts the foot and also holds up the arch.

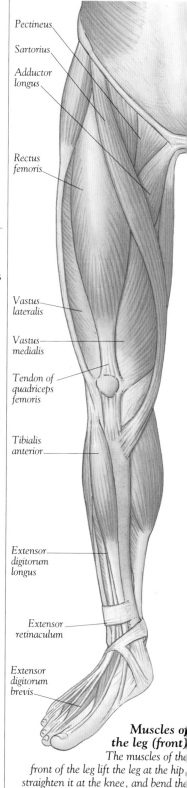

Pectineus

Sartorius

Adductor longus

Rectus femoris

Vastus lateralis

Vastus medialis

Tendon of quadriceps femoris

Tibialis anterior

Extensor digitorum longus

Extensor retinaculum

Extensor digitorum brevis

Muscles of the leg (front)
The muscles of the front of the leg lift the leg at the hip, straighten it at the knee, and bend the foot upward at the ankle

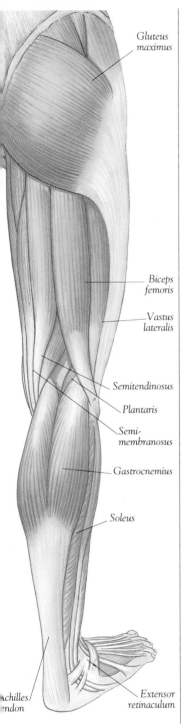

Gluteus
maximus

Biceps
femoris

Vastus
lateralis

Semitendinosus

Plantaris

Semi-
membranosus

Gastrocnemius

Soleus

Achilles
tendon

Extensor
retinaculum

Muscles of the leg (back)
*The muscles of the back of the leg
are concerned with turning and
straightening the leg at the hip, bending
it at the knee, lifting the heel upward,
and curling the toes downward.*

Gluteus maximus

A muscle that straightens the thigh

The gluteus maximus is the largest muscle in the body. It runs from the back of the pelvis to the upper part of the femur. You use this muscle when you stand up, walk, run, and climb stairs – in fact, whenever you straighten, or extend ■, your legs. Together with several other muscles, the gluteus maximus muscles form the **buttocks.**

Hamstring

A set of muscles that bends the knee

If you sit on a chair and tense your legs, you will feel strong tendons on the underside of your knees. These tendons belong to three muscles – the **biceps femoris**, the **semimembranosus** and the **semitendinosus**, which together are known as the hamstrings. They make the knee bend, or flex ■, and they also help to straighten the leg.

Flexor of the foot

A muscle in the lower leg that bends the foot or toes downward

When you stand on tiptoe, muscles in the back of your leg tighten up. These muscles are the flexors ■ of the foot, which make your foot and toes bend downward, away from the leg. There are nine of these muscles in the lower leg, arranged in two layers. The muscles in the surface layer bend the foot by pulling up the heel, and include the gastrocnemius. The muscles in the deep layer bend the toes by pulling on long tendons. They include the **flexor hallucis longus**, which bends the big toe, and the **flexor digitorum longus**, which bends the other toes.

See also

Arch 43 • Calcaneus 43 • Extend 46
Extensor 46 • Femur 41 • Flex 46
Flexor 46 • Interosseus muscle 55
Pelvis 41 • Tendon 49 • Tibia 41

Gastrocnemius

A flexor muscle that bends the foot downward

This muscle is the largest flexor of the foot. Its name means "belly of the leg," and its common name is the **calf muscle.** The gastrocnemius runs down the back of the lower leg, from the far end of the femur to the heel bone, or calcaneus ■. When it contracts, it makes the foot bend downward, and it also helps to bend the knee. The gastrocnemius is connected to the heel by the Achilles tendon.

Achilles tendon

An extremely strong tendon attached to the heel

The Achilles tendon is the strongest in the body. It is attached to the calcaneus and is pulled by three flexor muscles: the **soleus**, the **plantaris**, and the gastrocnemius.

*Stretching
tendons*
*This dancer's
Achilles tendons show clearly as
she stands on tiptoe and points her toes.*

Flexor digitorum brevis

A flexor muscle in the foot that bends the toes

This muscle on the underside of the foot flexes the small toes, while other muscles move the big toe. The foot also contains interosseus muscles ■ that enable the toes to spread out or close up.

Nerves

Nerves enable the different parts of your body to work as a single unit. They allow you to take in information about the world around you and to react quickly to changes. They also control many of your body's internal processes.

Nerve

A bundle of cells that carries signals

Nerves are the body's "wiring." Each one is covered by a tough, fibrous sheath and contains hundreds or thousands of cells called neurons. Nerves fan out from the brain ■ and spinal cord ■. They reach all parts of the body, including the skin, muscles, and sense organs ■, and even inside teeth and bones.

Neuron

A single nerve cell

Neurons are also known as **nerve cells**. They are adapted for carrying electrical signals, called nerve impulses. All neurons have a **cell body**, containing a nucleus ■, short filaments called **dendrites**, which carry electrical signals towards the cell body, and a long filament called an axon, which usually carries the signals away. Unlike most cells, neurons cannot divide ■ once formed. As the body grows, neurons that die are not replaced, and the number gradually decreases.

Synaptic knob

Axon

Myelin sheath formed by glial cells

Node of Ranvier

Dendrite

Cell body

Nucleus

Structure of a neuron
A typical neuron has a cell body, which contains the cell's nucleus, and a long axon. The axon is insulated by a myelin sheath. Between the glial cells that form the myelin sheaths are spaces called nodes of Ranvier.

Axon

A long filament in a neuron

An axon, or **nerve fiber**, is the part of a nerve cell that actually carries an electrical signal, or nerve impulse. Axons are finer than a hair, and although many are less than 0.04 inch (1 mm) long, some are over 3 ft (1 m) long. They are often surrounded by a sheath of a fatty substance called **myelin**. Myelin works like the plastic around an electrical wire, and helps to speed up the movement of nerve impulses.

Glial cell

A cell that supports, protects, or nourishes nerve cells

Glial cells do not carry nerve impulses. Instead, they support nerves by providing them with nutrients, or by attacking invading bacteria ■. Special glial cells called **Schwann cells** wrap themselves around the axons of some neurons and are rich in the insulating substance myelin.

Stimulus

A physical or chemical change that affects a nerve

A stimulus is anything that alters the electrical state of a neuron. If the stimulus is above a certain level, called a neuron's **threshold level**, it fires off a nerve impulse, producing a **response**. Stimuli are very varied. **External stimuli** come from outside the body and include changes in temperature, pressure, or light intensity. **Internal stimuli** come from inside the body. They include changes in osmotic pressure ■ and hormone levels.

Nerve impulse

signal that passes down a neuron

A resting neuron is rather like a charged battery. It uses an active transport ■ system called the **sodium-potassium pump** to move sodium ions ■ out through its plasma membrane ■, and to bring potassium ions in. This constant pumping creates a tiny electrical charge. If the cell receives a stimulus, sodium ions flood back across the membrane. This reverses the charge in that area and an electrical disturbance, or **action potential**, sweeps along the axon at up to 330 ft (100 m) per second. When the disturbance reaches a synapse, it can trigger an impulse in an adjacent nerve.

Synapse

a junction between two neurons

A synapse is a junction that allows one neuron to trigger an impulse in another. It consists of a small swelling at the end of an axon, called a **synaptic knob**, which is close to a neighboring neuron. When a nerve impulse reaches the synaptic knob, it triggers the release of a substance called a **neurotransmitter**. This crosses the gap between the two cells and in less than a millisecond, triggers the second neuron into action. Some neurons have hundreds or thousands of synapses. A synapse always passes a signal in the same direction; it cannot work in reverse.

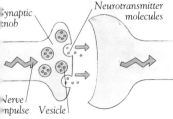

Synaptic knob — *Neurotransmitter molecules*

Nerve impulse — *Vesicle*

Synapse in action
Neurotransmitter molecules are released from tiny sacs called vesicles.

Neuromuscular junction

A special synapse between neurons and muscle fibres

A neuromuscular junction is a synapse between a motor neuron and a muscle fiber ■. When an impulse arrives at the synapse, it makes the muscle ■ contract.

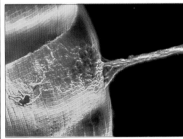

A neuromuscular junction
This false color electron micrograph shows a motor neuron (pink) attached to skeletal muscle fibers.

Sensory neuron

A neuron that carries signals to the central nervous system

Sensory neurons are also called **afferent neurons** (afferent means "carrying toward"). They carry nerve impulses from different parts of the body to the central nervous system (CNS) ■. Some sensory neurons are directly triggered by stimuli. Others are triggered indirectly, by special cells or neurons called **receptors**. Some receptors are scattered throughout the body, others are clustered in special sense organs, such as the eyes and ears.

Motor neuron

A neuron that stimulates an effector

Motor neurons are also known as **efferent neurons** (efferent means "carrying away"). They carry nerve impulses away from the CNS to structures called **effectors**. The most important effectors are muscles and glands ■.

Camillo Golgi

Italian histologist (1843–1926)

Camillo Golgi made an important breakthrough that allowed nerve cells to be seen clearly for the first time. He devised a technique that stains nerve cells black, but leaves the surrounding cells almost unaffected. Using this stain, he identified different types of nerve cell, and found that synapses contain small gaps between one cell and the next. He also discovered the Golgi body ■, an organelle ■ that is present in most cells.

Association neuron

A neuron that passes signals from one neuron to another

Association neurons enable the nervous system to pass on signals, and to sort and compare them. These neurons are found mainly in the central nervous system.

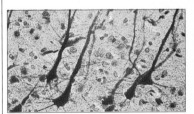

Brain power
This light micrograph shows association neurons in the brain.

Nervous system

Billions of interconnnected neurons extend throughout your body, making up its communication network, or nervous system. Together, they collect, process, and distribute all the information your body needs to work.

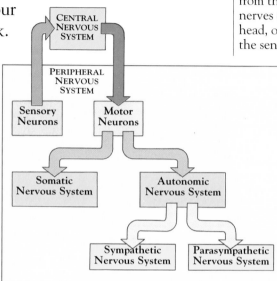

Nervous system

A network of neurons that extends throughout the body

The nervous system controls and coordinates the activities of the entire body. It is divided into two parts. The **central nervous system**, or **CNS**, consists of the spinal cord ■ and the brain. It analyzes incoming information, stores it, and issues instructions. The **peripheral nervous system** consists of **peripheral nerves** that transmit signals between the CNS and the rest of the body.

Somatic nervous system

A part of the nervous system that controls voluntary actions

Whenever you make a conscious movement, your somatic nervous system is at work. It carries signals to the skeletal muscles ■, and makes them contract. The somatic nervous system extends throughout the body, and is part of the peripheral nervous system.

See also

Gland 78 • Heart rate 87 • Humerus 41
Neuron 58 • Olfactory bulb 75
Optic chiasma 71 • Radius 41 • Receptor 59
Skeletal muscle 48 • Smooth muscle 49
Spinal cord 62 • Tibia 41 • Vertebra 36

The nervous system
Sensory neurons gather information from the body and pass it to the CNS. The CNS analyzes this information and relays instructions back to the peripheral nervous system.

Autonomic nervous system

A part of the nervous system that controls involuntary actions

The autonomic nervous system is part of the peripheral nervous system. It regulates processes in the body that seem automatic. The autonomic nervous system dilates pupils, controls sweating, stimulates salivation, regulates heart rate ■ and blood pressure, and controls a variety of glands ■. It is divided into two parts – the **sympathetic nervous system** and the **parasympathetic nervous system**. These produce opposite effects, and together, they help to keep the body in a stable state.

Cranial nerve

A peripheral nerve arising from the brain

Peripheral nerves fan out from two regions – the brain and the spinal cord. There are 12 pairs o cranial nerves, which emerge from the brain. Most of these nerves control muscles in the head, or carry information from the sense organs to the brain.

Olfactory nerve

A cranial nerve used in the sense of smell

Olfactory nerves carry signals that the brain interprets as smells. These nerves form two swellings, called olfactory bulbs ■, that lie just beneath the front of the brain.

Facial nerve

A cranial nerve used to control facial expression

The facial nerve extends over the face and neck. Like many nerves, it contains two types of neuron ■. Motor neurons contro the muscles that produce facial expressions, and sensory neurons carry messages back to the brain.

Optic nerve

A cranial nerve used in vision

Optic nerves carry signals from the eyes to the brain. The brain uses these signals to build up a picture. Optic nerves do not run directly to the brain. Instead, they partly cross over in a structure called the optic chiasma ■. This allows each side of the brain to deal with just hal of the image that you see. Three other cranial nerves control the movement of the eyeball, and control the change in shape of the lens and the pupil.

Spinal nerve

A peripheral nerve arising from the spinal cord

There are 31 pairs of spinal nerves, each of which emerges from gaps between the vertebrae ■ in the spinal cord. Spinal nerves are divided into four groups – **cervical nerves**, **thoracic nerves**, **lumbar nerves**, and **sacral nerves**. The nerves that emerge from the middle of the spinal cord spread outward to reach receptors ■, muscles, and glands in the trunk. The nerves that emerge from the top and bottom of the spinal cord join to form special networks called **plexuses**. The main plexuses are the **cervical plexus**, the **brachial plexus**, the **lumbar plexus**, and the **sacral plexus**. Nerves from these plexuses control the arms and legs.

Sciatic nerve

A nerve that controls muscles in the leg and foot

The sciatic nerve arises from the sacral plexus and is the largest nerve in the body. It starts near the base of the spinal cord, and runs through the thigh before dividing into two branches above and behind the knee. One branch, the **tibial nerve**, runs down the back of the lower leg close to the tibia ■. The other branch becomes the two **peroneal nerves**, which run down the front of the lower leg. The sciatic nerve controls many of the muscles used for standing and moving about.

Brain
Cranial nerves
Cervical nerves
Spinal cord
Brachial plexus
Thoracic nerves
Median nerve
Radial nerve
Ulnar nerve
Lumbar plexus
Sacral plexus
Lumbar nerves
Sacral nerves
Sciatic nerve
Common peroneal nerve
Tibial nerve
Deep peroneal nerve

The nervous system
The central nervous system (CNS) consists of the brain and spinal cord, while the peripheral nervous system is made up of a network of nerves that spreads throughout the body.

Radial nerve

A nerve that controls muscles on the outer margin of the arm and hand

The radial nerve is one of three nerves that control parts of the arm and hand. It runs the length of the arm, and lies close to the radius ■ in the forearm, hence its name. The radial nerve controls many of the muscles that straighten the hand and fingers. The **median nerve** runs down the center of the forearm, and makes the wrist and fingers bend. Both of these nerves arise from the brachial plexus.

Ulnar nerve

A nerve that controls muscles on the inner margin of the arm and hand

The ulnar nerve arises from the brachial plexus. It runs down the arm and passes through a hollow in the back of the elbow. Here, the nerve lies just under the skin and is easily caught during sudden movements. This area is often called the **funny bone**, because the nerve passes over the humerus ■. The ulnar nerve carries sensations from the little finger, so this finger often tingles if you hit your funny bone.

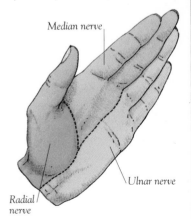

Median nerve
Ulnar nerve
Radial nerve

Nerves of the hand
Each of the three nerves that control the hand affects a well-defined area.

Spinal cord

The spinal cord is the "highway" of the body's communication network. It relays information from the body to the brain, and sends instructions to other areas of the body. Through reflexes, it allows you to react quickly to hazards and danger.

Spinal cord

A column of nervous tissue that runs through the backbone

The spinal cord is an extension of the brain ■. It relays messages to the brain and is also involved in many reflexes. The spinal cord runs from the base of the skull ■ to a point just over halfway down the back and is protected by a bony tube of vertebrae ■. It varies in width but at its midpoint, it is about as thick as a finger. The spinal cord contains huge numbers of nerve cells or neurons ■, and gives rise to the paired spinal nerves ■ which branch throughout the body.

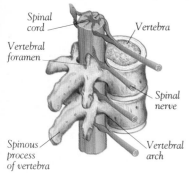

The vertebral column
The spinal cord passes through a space called the vertebral foramen and is protected by the vertebral arch.

See also

Association neuron 59 • Axon 58
Brain 64 • Coccyx 37 • Glial cell 58
Ligament 44 • Motor neuron 59
Neuron 58 • Sensory neuron 59
Skull 38 • Spinal nerve 61 • Vertebra 36

Gray matter

A tissue consisting mainly of cell bodies of neurons

The spinal cord is roughly oval in cross-section. At its center is a dark H-shaped mass made of gray matter. Gray matter relays information between the spinal cord and spinal nerves. It consists mainly of the cell bodies of neurons, together with glial cells ■ and blood vessels.

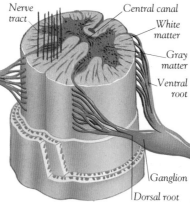

Structure of the spinal cord
The spinal cord consists of a central core of gray matter surrounded by a tube of white matter.

White matter

A tissue consisting mainly of axons

White matter makes up the outer part of the spinal column. It consists mainly of axons ■, and it carries signals up and down the spinal column, and to and from the brain. The axons in white matter are surrounded by myelin, which gives the white matter its white color.

Cerebrospinal fluid

A liquid that nourishes and protect the spinal cord and brain

The entire central nervous system floats in a jacket of clear cerebrospinal fluid, or **CSF**. This fluid has two functions. It works like a shock absorber, protecting the spinal cord and brain from sudden jolts. It also delivers nutrients collected from the blood and carries away waste products. An adult has about a cupful of CSF within the central nervous system. It flows around the spinal cord and the brain in a cavity called the **subarachnoid space**. It also flows through the spinal cord in a narrow channel called the **central canal**.

Nerve root

The point where a nerve meets the spinal cord

Before a nerve merges with the spinal cord, it splits into two branches, or roots. The root nearest the front of the body is called the **ventral root**, or **motor root**. This root contains motor neurons ■, which carry signals from the spinal cord to muscles. The root nearest the back of the body is the **dorsal root**, or **sensory root**. It is made up of sensory neurons ■, which bring signals to the spinal cord from the body. The dorsal root has a swelling called a **ganglion** which contains the cell bodies of the sensory neurons.

Nerve tract

A bundle of neurons in the spinal cord or brain

A nerve tract is a communication link inside the central nervous system. It is made up of neurons that work together to carry signals from one place to another Tracts run through the spinal cord, and also through the brain.

The spinal cord and spinal nerves

The spinal cord runs through a bony tunnel of vertebrae. Pairs of spinal nerves leave the spinal cord at gaps between the vertebra. In adults, the cord ends just below the 12th rib, where the cauda equina begins.

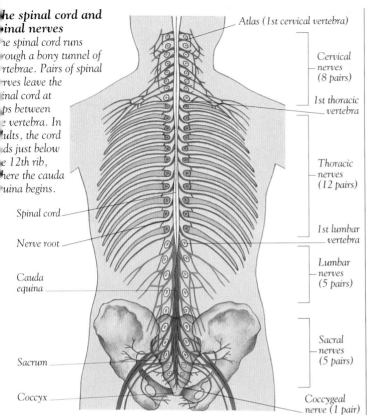

Spinal cord

Nerve root

Cauda equina

Sacrum

Coccyx

Atlas (1st cervical vertebra)

Cervical nerves (8 pairs)

1st thoracic vertebra

Thoracic nerves (12 pairs)

1st lumbar vertebra

Lumbar nerves (5 pairs)

Sacral nerves (5 pairs)

Coccygeal nerve (1 pair)

Stretch reflex

A reflex that counteracts a sudden pull on a muscle

If one of your muscles is suddenly stretched, it tightens up. This slows down the stretching, and helps to prevent injury. The **knee-jerk reflex**, or **patellar reflex**, is a stretch reflex. To test it, a doctor taps the ligament ▪ below the kneecap. This pulls the kneecap down and briefly stretches muscles in the thigh. If the reflex is working, the thigh muscles suddenly contract, and the lower leg kicks upward.

Inborn reflex

A reflex present at birth

Everybody is born with a set of reflexes that are built into the body's nervous system. Some of these reflexes are only seen in infants; others stay with us throughout life. A **conditioned reflex** is one that is learned by everyday experience or repetition.

Cauda equina

A collection of nerves at the base of the spinal cord

Cauda equina means "horse's tail." It describes a fan-shaped collection of nerves that begins at the end of the spinal cord. As pairs of nerves branch off, the remaining collection of nerves becomes narrower. Eventually, all that is left is a thin filament of tissue attached to the coccyx ▪.

Reflex

A rapid response to a stimulus

A reflex is a split-second reponse that protects you from danger or everyday hazards. During a reflex action, your body reacts without waiting for you to think. Reflexes are usually triggered by simple nervous pathways called reflex arcs. Many of these involve the spinal cord, but not the brain.

Reflex arc

A nervous pathway involved in a reflex

Reflexes have to be fast, so their signals travel along the shortest pathways. For example, if you tread on a sharp object, a sensory neuron sends a signal to your spinal cord. An **association neuron** passes the signal to a motor neuron, and your foot is moved. Only later do signals reach the brain.

Primitive reflex

A reflex that pulls the body away from danger

If you touch something hot or painful, this life-saving reflex pulls your hand away. The withdrawal reflex is shown by many body parts, but is most noticeable in the hands and feet.

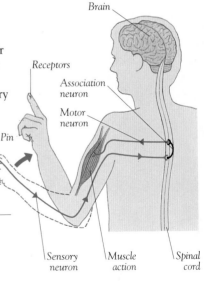

Brain

Receptors

Association neuron

Motor neuron

Pin

Sensory neuron

Muscle action

Spinal cord

Withdrawal reflex
Receptors in the hand detect pain caused by a pinprick, and send signals to the spinal cord via a sensory neuron. Instructions are sent back via a motor neuron to activate a muscle.

Brain

The human brain is the control center for the body and contains billions of nerve cells. When they are working hard, these cells use about 20 percent of the body's oxygen. The brain is organized into several "departments," each running a different part of the body.

Forebrain

A part of the brain that deals with homeostasis, emotions, and conscious actions

Most of the forebrain consists of the cerebrum, which is where conscious thought takes place. Hidden beneath the cerebrum are two much smaller structures the hypothalamus and thalamus

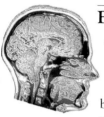

Computed tomograph of the brain

Brain

An organ that processes information

A human brain can weigh up to 3lb (1.4 kg), and is one of the largest organs in the body. It is divided into three regions: the brain stem, the cerebellum, and the forebrain. Like the spinal cord ■, the brain consists mainly of gray matter ■ and white matter ■. These are arranged in distinct layers.

Brain stem

The part of the brain that merges with the spinal cord

The brain stem is the body's equivalent of an autopilot. It controls the basic processes that are essential for life, such as breathing, rate of heartbeat, and blood pressure. The lowest part of the brain stem is the **medulla oblongata**, or **medulla**. The medulla maintains these vital functions, and is the point where many nerve fibers cross over. Higher up is a swelling called the **pons**, which means "bridge." The pons relays information between the brain and the spinal cord. The highest part of the brain stem is the **midbrain**, which is involved in many reflexes ■.

Cerebellum

The part of the brain that coordinates subconscious movements

The cerebellum ensures that your body is balanced and moves in a coordinated way. It consists of millions of neurons ■ packed into two folded halves, or **hemispheres**, and makes up about 10 percent of the brain's weight. The cerebellum constantly receives "updates" about the body's position and movements. By sending instructions to muscles, it adjusts the body's posture and keeps it moving smoothly.

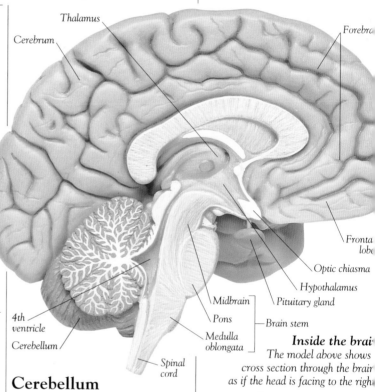

Thalamus

Cerebrum

Forebra

Fronta lobe

Optic chiasma

Hypothalamus

Pituitary gland

Midbrain

Pons

Medulla oblongata

Brain stem

4th ventricle

Cerebellum

Spinal cord

Inside the brai
The model above shows cross section through the brain as if the head is facing to the right

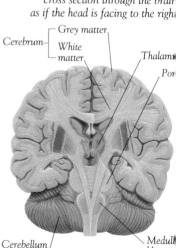

Cerebrum—Grey matter, White matter

Thalam

Por

Cerebellum

Medull oblongat

Frontal section through brain

Cerebrum

The part of the brain involved in emotions and conscious thought

The cerebrum accounts for about 85 percent of the brain's weight. It has two halves, called **cerebral hemispheres**. These consist of folds called **gyri** (singular **gyrus**), small grooves called **sulci** (singular **sulcus**), and deeper grooves called **fissures**. The surface layer of each hemisphere is made up of gray matter, and is known as the **cerebral cortex**. This is where the cerebrum processes information. Beneath the cortex is a middle layer of white matter and deep areas of gray matter called **basal ganglia**. Nerve tracts ■ link the different parts of each hemisphere and also connect the two hemispheres.

Sensory area

A region of the cerebral cortex that analyses information from the sensory receptors

Sensory areas receive information from sense organs ■ and other receptors ■ throughout the body. They sort and analyze the information so it can be understood. Information from different senses is managed by different parts of the cortex. For example, signals from the eyes are dealt with initially by the optic lobe, or **visual cortex**, which is at the back of the brain.

Motor area

A region of the cerebral cortex that controls voluntary movement

Motor areas control the body's skeletal muscles ■. Motor nerves cross as they leave the brain, so each motor area lies on the opposite side of the body to the region that it controls. If a motor area is damaged – for example, by a stroke ■ – that part of the body ceases to move normally.

Association area

A part of the cerebral cortex involved in thought and comprehension

Association areas make us fully conscious ■ and give us awareness. They allow us to analyze experiences, and look at things in a logical or artistic way. Unlike sensory and motor areas, association areas are not the same on both sides of the brain. In many people, the left hemisphere deals with logic and comprehension, while the right one deals with the perception of shapes and feelings.

Touch sensation (A)   Movement (S) (basic) (M)   Movement (skilled) (A)
Visual recognition (A)   Thought (A)
Speech (M)
Vision (S)   Sound (A)   Sound (S)   Taste (S)

Areas of the cerebral cortex
The cerebral cortex has motor (M), sensory (S), and association (A) areas.

Hypothalamus

A region of the forebrain that monitors the state of the body

The hypothalamus is a small but important region that is located below the thalamus. It monitors factors such as body temperature and food intake, and issues instructions that correct any imbalances. The hypothalamus works partly by sending signals through the autonomic nervous system ■, and partly by triggering the nearby pituitary gland ■ to release hormones ■.

Thalamus

A region of the forebrain that relays sensory information to the cerebrum

The word thalamus means "inner room," and it describes a small, round cluster of nerve cells deep inside each cerebral hemisphere. The thalamus sends information from the sense organs to the sensory areas of the cerebral cortex, and sends motor signals in the opposite direction.

Ventricle

A fluid-filled cavity in the brain

The brain is not entirely solid. It contains four ventricles which connect with each other, and with the space around the brain and the spinal cord. Ventricles produce cerebrospinal fluid ■, a clear, shock-absorbing liquid that is constantly moving.

Meninx

A protective membrane around the brain and spinal cord

The entire central nervous system is protected by three membranes, called the **meninges** (the plural of meninx). The outer membrane, the **dura mater**, is attached to the inside of the skull and forms a tube around the spinal cord. Inside this is the **arachnoid layer**, which has a weblike network of cells. The inner membrane, the **pia mater**, has a rich blood supply and is attached to the surface of the brain and spinal cord.

See also

Brain & behavior

Every moment that you are awake, you are interacting with the world around you. Your brain perceives sensations from inside and outside your body, and it processes and reacts to information to make you behave in a certain way.

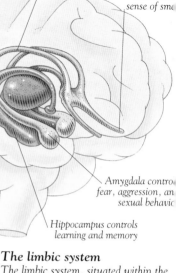

Limbic cortex modifies behavior

Olfactory bulb controls the sense of smell

Amygdala controls fear, aggression, and sexual behavior

Hippocampus controls learning and memory

Consciousness

Awareness of self and surroundings

As far as we know, humans are the only living things that are fully aware, or **conscious**, of their own existence. Consciousness is produced by the cerebrum ■ and is the highest level of mental activity. It is accompanied by **subconscious** mental activity, which occurs without our direct knowledge that it is taking place. During **unconsciousness**, the body continues to function, but the brain is not aware of its surroundings, and does not respond to them. A **coma** is a state of deep unconsciousness that may be produced by an accident or illness.

Intelligence

The ability to analyze and reason

Intelligence is a collection of abilities – being able to learn, to remember, to understand and to solve problems. Someone who scores well on exams is not necessarily very intelligent – he or she may just have a good memory and good test-taking skills.

See also

Body temperature 76 • Cerebrum 65
Forebrain 64 • Hypothalamus 65
Inherited characteristic 134
Nerve impulse 59 • Smell 75

Behavior

A pattern of responses to the outside world

Many animals often behave in a very predictable way, because they work entirely by inherited "instructions" called **instincts**. Humans also show **instinctive behavior**, or **innate behavior**, particularly when newly born. However, a large amount of our behavior develops in another way – by **learning**. Unlike most animals, humans take a long time to grow up. During that time, we learn how to interact with each other, and develop skills that will be useful in later life.

Learning and reasoning
The human brain is capable of matching information that it has learned in the past with new situations or problems. This is called reasoning. Faced with a new puzzle, this child works it out by recognizing the shapes and remembering puzzles that she has done before.

The limbic system
The limbic system, situated within the brain, controls emotions and behavior.

Emotion

A state of mind

Pleasure, disappointment, hope, and anger are all emotions. Emotions are usually triggered by events that we experience, but they can be altered by other factors, such as drugs. Emotions are controlled by part of the forebrain ■ called the **limbic system**, which lies at the base of the cerebrum. The limbic system also perceives smell ■ and controls instinctive behavior.

Memory

The ability to remember

The brain is capable of absorbing and storing a huge amount of information. Exactly how and where it does this is still not understood. But scientists do know that there are two main kinds of memory. Your **short-term memory** stores most, or perhaps all, of what you experience, but only for a brief period. Your **long-term memory** holds selected information only, but can store it for many years.

Handedness

The tendency to use one hand more than the other

About 90 percent of people are right-handed, which means that they use their right hand for anything that involves careful coordination. The remaining 10 percent are either left-handed, or ambidextrous, which means they can use their right and left hand equally well. Handedness is controlled by the cerebrum in the brain. It is an inherited characteristic ■, but it can be altered by learning. At one time, left-handed children were forced to become right-handed. Today, most people realize that it does not matter which hand you use.

Even contest?
Left- and right-handed players play squash together on the same court.

Circadian rhythm

A cycle in the body that is about 24 hours long

Many of the body's activities follow circadian rhythms. People sleep for roughly the same period every 24 hours. Blood pressure and body temperature ■ also rise and fall in a 24-hour cycle. A chemical mechanism called a biological clock, thought to be located in the hypothalamus ■, controls these internal rhythms.

Sleep

A state of altered consciousness during rest

Sleep allows the body to rest and lets the brain "recharge its batteries." Sleep is not the same as unconciousness, because a sleeping person can wake up. Normal sleep follows a pattern of cycles. Each cycle starts with a period of **nonrapid eye movement (NREM) sleep**, which is also known as **deep sleep**. During this stage, the brain and eyes are fairly inactive but the body may move about. After 90–120 minutes, there is a sudden burst of **rapid eye movement (REM) sleep**. This is also known as **light sleep**. During this stage, the eyes dart about and the eyelids flicker. The brain is highly active, but the body remains still. REM lasts for 10–60 minutes, before the cycle begins again.

Dreaming

A sequence of events imagined during sleep

Dreams occur during REM sleep, when the sleeping brain is most active. During a dream, the brain links objects and experiences in an apparently jumbled way and it often mixes up past and present events. Some scientists think that dreams are the result of the brain working at random. Others believe that dreaming helps the memory.

Sleepwalking

Walking while remaining asleep

Sleepwalking occurs during NREM sleep, when the brain is fairly inactive. A sleepwalking person moves slowly, and often seems unaware of people around them. Sleepwalking is common in children, but rare in adults.

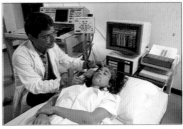

The brain on tape
To record brainwaves, electrodes attached to the patient's head are connected to an electroencephalograph.

Brainwave

An electrical wave created by neurons in the brain

The brain works by handling millions of nerve impulses ■ every second. These impulses produce an **electrical field** around the head. The strength of the field can be measured with an instrument called an **electroencephalograph** or **EEG**, which produces a reading called an **electroencephalogram**. This shows that the field rises and falls in cycles called brainwaves. There are four types of brainwave. **Alpha waves** occur when someone is awake, and **delta waves** occur during NREM sleep. **Beta waves** and **theta waves** occur when the brain is very active.

Alpha waves

Beta waves

Theta waves

Delta waves

Brainwaves
Brain waves can be recorded to check if the brain is working normally and, if not, to diagnose brain diseases.

Vision

For most people, vision is the most important of the five "special senses." We use sight to create images of our surroundings. These images, made by signals from the eyes, help us to move about and to communicate.

Sense

A system that detects specific changes inside or outside the body

Senses allow you to respond to the world around you, and to changes inside your body. Humans have five **special senses**: vision, hearing, balance, taste, and smell. These senses work through receptors ■ located in special **sense organs**. There are also **general senses**, such as touch, that work via receptors scattered throughout the body.

Vision

A sense that detects light

In vision, or **sight**, light energy from an object is gathered and then focused by the eye to form a picture, or **image**. The image falls on cells called photoreceptors, which send signals to the brain. Together, these signals convey a huge amount of information.

Eye

A sense organ that detects light

The eye consists of a spherical **eyeball** about 1 inch (2.5 cm) across, which sits in a bony socket called the orbit ■. The surface of the eyeball has three layers. The outer layer, the **sclera**, makes up the white of the eye and is tough and slippery. The central layer, the **choroid**, supplies the eye with blood. The inner layer, or retina, contains light-sensitive cells. The eyeball is divided into two unequal spaces by a flexible lens.

Cornea

A transparent layer at the front of the eyeball

The cornea protects the front of the eye and helps to focus light. It is covered by a membrane called the **conjunctiva**, which also lines the eyelids. The conjunctiva is bathed by **tears**, which lubricate the eye. Tears contain the substance lysozyme, which helps to kill bacteria ■ and thus stops infection. They are made by **lacrimal glands**. After flowing over the eye, tears drain into the **lacrimal sac** inside the nose.

Iris

A ring of muscular tissue between the cornea and the lens

The iris alters the size of the **pupil**, which is the hole that lets light into the eye. It contains smooth muscle ■ and works by an automatic reflex ■. The iris makes the pupil close up in bright light and widen when it is dim. It also contains pigments that give the eyes their color.

Dilated pupil
In bright light, the pupil contracts (left) to prevent too much light from entering the eye. This helps to protect the nerve cells at the back of the eye. In dim light, the pupil dilates (right) to let in as much light as possible.

Lens

A curved structure that forms an image by bending light rays

Each eye has a single lens, which is made of a transparent protein. The shape of the lens can be changed by the ring-shaped **ciliary muscle**. This allows the eye to bring everything you see into focus by a process called accommodation ■. An upside-down image is formed on the retina by the lens. It is like this from the moment you are born, so you are unaware that everything you see is inverted. The lens becomes less flexible with age, and this often causes a visual defect called presbyopia ■.

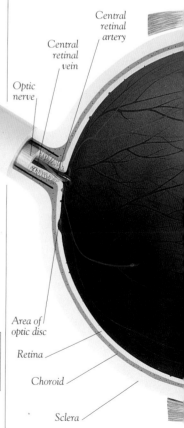

Central retinal artery

Central retinal vein

Optic nerve

Area of optic disc

Retina

Choroid

Sclera

Section through the eye
This model shows a cross-section of the right eye, seen from above.

Retina

A membrane containing cells that detect light

The retina is a thin screen that lines the rear half of the eye. It has a rich blood supply and is packed with about 125 million photoreceptors. These light-sensitive cells are connected to the optic nerve ■, along which they send signals to the brain. The central point of the retina, called the **fovea**, is particularly good at sensing detail in bright conditions. However, it does not work very well in dim light.

Retina and blind spot
The blind spot is shown as a light circular area, bottom right, on this view of the retina.

— Optic disc

Blind spot

The region where the optic nerve leaves the retina

The blind spot is a small part of the retina that cannot detect light. It marks the point where the optic nerve leaves the eyeball at the **optic disc** on its way to the brain. Normally, you are not aware of your blind spot because your brain ignores the small "gap" that it creates.

Photoreceptor

A cell that responds to light

Photoreceptors contain a colored chemical, or **visual pigment**, that changes shape when it is struck by light. This change triggers a nervous impulse, which travels to the brain. The eyes have two kinds of photoreceptor – **rods** and **cones**. Rods are more numerous than cones, and contain the pigment **rhodopsin**. They work well in dim light, but only "see" in black and white. Cones have similar pigments. They give you color vision ■.

See also

Accommodation 70 • Bacteria 92
Color vision 71 • Farsightedness 70
Optic nerve 60 • Orbit 39 • Receptor 59
Reflex 63 • Smooth muscle 49

Vitreous humor

A transparent substance that fills most of the eye

Also known as the **vitreous body**, this jellylike substance fills the eye behind the lens. It gives the eye its shape, and keeps the retina in position. The part of the eye in front of the lens is filled with a watery liquid called the **aqueous humor**. This is under slight pressure, and keeps the cornea curved outward.

Eyelid

A flap of skin that protects and cleans the eye

Your eyelids protect your eyes in several ways. They shut almost instantly if anything heads toward your eyes. They also shut when you are asleep, and in very bright light. When you **blink**, your eyelids wipe your eyes and keep them free of dust and other particles. In bright conditions, **eyelashes** and **eyebrows** stop too much light from entering the eye.

Retinal blood vessel
Vitreous humor
Ciliary muscle
Iris
Conjunctiva
Pupil
Cornea
Aqueous humor
Lens

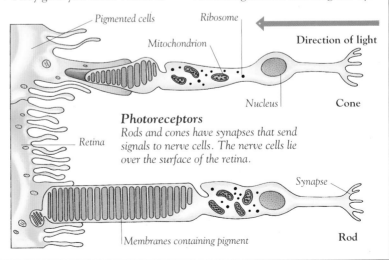

Pigmented cells
Ribosome
Mitochondrion
Direction of light
Nucleus
Cone
Retina
Synapse
Rod
Membranes containing pigment

Photoreceptors
Rods and cones have synapses that send signals to nerve cells. The nerve cells lie over the surface of the retina.

Continued over page ➤

Defects of vision

Most common vision defects are caused by misshapen parts of the eye, such as the lens, cornea, or eyeball. Aging can also make the eyes become less flexible.

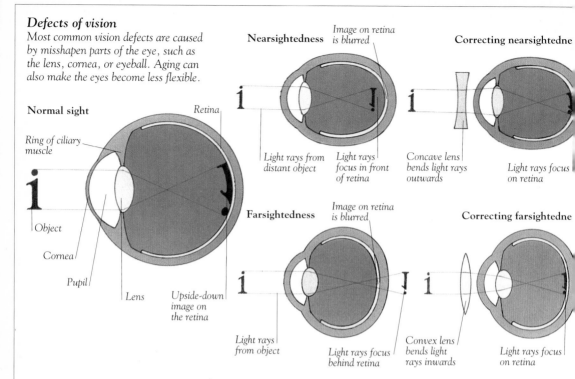

Normal sight

Ring of ciliary muscle

Retina

Object

Cornea

Pupil

Lens

Upside-down image on the retina

Nearsightedness

Image on retina is blurred

Light rays from distant object

Light rays focus in front of retina

Correcting nearsightedne

Concave lens bends light rays outwards

Light rays focus on retina

Farsightedness

Image on retina is blurred

Light rays from object

Light rays focus behind retina

Correcting farsightedne

Convex lens bends light rays inwards

Light rays focus on retina

Accommodation

Focusing brought about by a change in the shape of the lens

When light enters your eyes, it is bent, or **refracted**, by the cornea ▪ and the lens ▪. The cornea is fixed in shape, but muscles in the eye can change the shape of the lens. Such changes are called accommodation. If you look at a distant scene, the light rays that enter your eye are almost parallel. The lens remains thin and flat, as it does not have to bend the rays very far. If you look at a close object, the light rays that enter your eye diverge. The ciliary muscle ▪ contracts to make the lens thicker and more curved, so that the rays can be brought into focus.

See also

Ciliary muscle 68 • Cone 69 • Cornea 68
Eyeball 68 • Lens 68 • Optic nerve 60
Photoreceptor 69 • Retina 69
Sex-linked allele 135 • Visual pigment 69

Nearsightedness

A defect of vision in which light is focused in front of the retina

Nearsightedness is also known as **myopia**. A nearsighted person is unable to see distant objects well because their eyes focus light before it reaches the retina ▪. The result is a blurred image on the retina. Nearsightedness is usually caused by the eyeball ▪ being too long or the lens being too thick. It affects people of all ages, and can be corrected by wearing glasses or contact lenses.

Astigmatism

A defect of vision caused by an irregularly curved cornea

A person with astigmatism cannot focus equally across their whole field of view. Some areas appear sharp, but others are blurred. Astigmatism is caused by a cornea that does not have a symmetrically curved surface. It can be corrected by wearing glasses or contact lenses.

Farsightedness

A defect of vision in which light is focused behind the retina

A person with **farsightedness** is unable to see close objects clearly, because their eyes focus light behind the retina, rather than on it. There are two kinds of farsightedness. In **hyperopia**, or **hypermetropia**, the eyeball is too short in relation to the lens. The lens cannot focus light on nearby objects, so they look blurred. Hyperopia affects peopl of all ages, but can be corrected by wearing glasses. In **presbyopia** which usually develops from abou 45 years onwards, the eyeball is the right size, but the lens begin to lose its ability to focus. The ey can be focused on a point in the distance, so that distant objects can be seen fairly well, but near objects cannot. Presbyopia is corrected by wearing glasses for close work, such as reading. Som older people may need **bifocal** glasses, with separate lenses for close and distant objects.

◄ Continued from previous page

Color vision

The ability to distinguish different wavelengths of light

Light energy travels in waves, and the distance between one wave and the next gives the light its color. The eyes use cells called cones ■ to distinguish different colors. Cones contain a type of visual pigment ■, which exists in three different forms. This allows cones to respond to either red, green, or blue wavelengths of light. The brain combines signals from the cones to see a color image. Cones work only in bright light, so you cannot distinguish colors when it is dark.

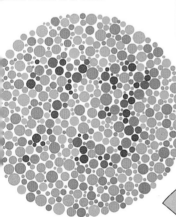

Color test
A person with red-green color blindness is unable to see this number.

Color blindness

An inability to distinguish certain colors

A person who is color blind, or **color deficient**, cannot tell some colors apart. This is because their eyes lack some of the cones that are used in color vision. The most common form of this defect, **red-green color blindness**, is inherited, and it is also sex-linked ■. It is common in males, but rare in females. Complete color blindness, in which the eyes can only see in black and white, is rare.

Binocular vision

Vision that uses two eyes

Each eye sees from a slightly different position. The brain compares the signals from both eyes, and uses this information to judge how far away an object is. Normally, both eyes look in the same direction, so their fields of view overlap by a set amount. In a person who develops **strabismus**, or crossed eyes, the eyes point in different directions. If this happens, the brain often ignores the signals from one of the eyes.

Optic chiasma

A cross-over point between the two optic nerves

Each eye is linked to the brain by a single optic nerve ■. The two nerves partly cross at a junction called the optic chiasma. Signals from the right halves of both eyes go to the right half of the brain, and signals from the left halves go to the left of the brain.

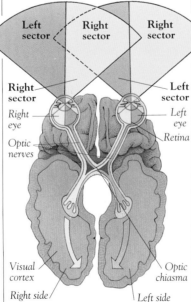

Optic chiasma
This view of the brain from below shows the optic chiasma, where the optic nerves partly cross.

Thomas Young

English physiologist (1773–1829)

Thomas Young was interested in all aspects of light, including how we perceive it. In 1801, he suggested that eyes see colors by responding to just three different wavelengths. According to his theory, the brain monitors the mix of these three wavelengths, and uses this information to see a full-color image. Young's idea of "trichromacy" turned out to be correct. This system is used in the eye, and also in color printing and color television.

Dark adaptation

A change in the eye brought about by low light levels

If you walk into a movie theater, you may have trouble finding your way to a seat. After about 15 minutes, however, you will be able to see easily. This is because your eyes adapt to the dark. The photoreceptors ■ in your eyes increase the amount of visual pigment. This makes them more sensitive and enables you to see in the dark. In bright light, the process goes into reverse, and the photoreceptors become less sensitive.

Night blindness

An inability to see normally in low light levels

A person with night blindness has difficulty adjusting to low levels of light, so they cannot see very well in the dark. Night blindness can sometimes be caused by a lack of vitamin A, which is needed to make the visual pigment rhodopsin.

Hearing

When an aircraft roars overhead, or someone whispers in your ear, air begins to vibrate. When the vibrations reach your ear, they trigger off a chain of movements in your skull. The result is what you perceive as sound.

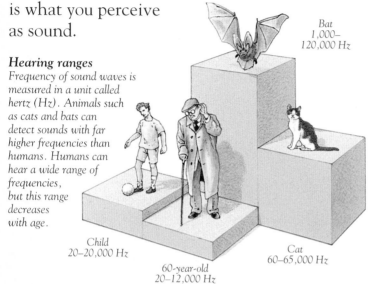

Bat
1,000–
120,000 Hz

Child
20–20,000 Hz

60-year-old
20–12,000 Hz

Cat
60–65,000 Hz

Hearing ranges
Frequency of sound waves is measured in a unit called hertz (Hz). Animals such as cats and bats can detect sounds with far higher frequencies than humans. Humans can hear a wide range of frequencies, but this range decreases with age.

Eardrum
A membrane that is vibrated by sound waves

The eardrum is also known as the **tympanic membrane**. This thin, taut circle of tissue is found between the end of the auditory canal and the middle ear. When sound waves are channeled into the ear the eardrum vibrates. These vibrations are then carried through the middle ear to the inner ear, so that they can be sensed. The eardrum is a delicate object, and can be burst, or **perforated**, by sudden changes in air pressure, or by infections. If this happens, hearing is affected until the eardrum heals.

Structure of the ear
Sound vibrations travel through air within the outer ear. They are passed on by movement of the bones inside the middle ear until they reach the inner ear, where they pass through fluid.

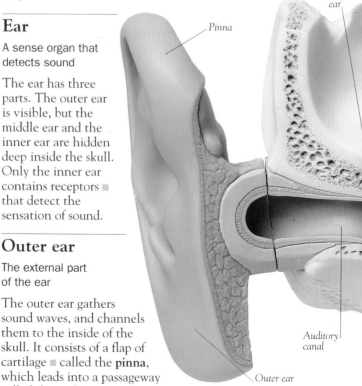

Pinna

Middle ear

Auditory canal

Outer ear

Hearing
A sense that detects sound

Sound is created by waves of pressure that travel through air, liquids, and solids. The **loudness** of a sound depends on the strength of its waves. The quality, or **pitch**, of a sound depends on the **frequency** of the waves. Frequency is a measure of how quickly one wave is followed by the next. Most people can hear sounds with a frequency of between 20 and 20,000 hertz (waves per second). As you get older, your ability to hear high-pitched sounds is reduced.

See also

Ear
A sense organ that detects sound

The ear has three parts. The outer ear is visible, but the middle ear and the inner ear are hidden deep inside the skull. Only the inner ear contains receptors ■ that detect the sensation of sound.

Outer ear
The external part of the ear

The outer ear gathers sound waves, and channels them to the inside of the skull. It consists of a flap of cartilage ■ called the **pinna**, which leads into a passageway called the **auditory canal**.

Middle ear

An air-filled space in the skull that contains the ear bones

In the middle ear, three tiny bones called **ear ossicles** carry vibrations from the eardrum to the inner ear. The bones are named after their shape. The **malleus**, or **hammer**, is nearest the eardrum, followed by the **incus**, or **anvil**, and finally the **stapes**, or **stirrup**. The stapes is 0.2 inch (5 mm) long and is the smallest bone in the body.

Eustachian tube

A tube that connects the middle ear with the throat

The Eustachian tube allows air into the middle ear, so that the pressure on both sides of the eardrum is kept equal. The tube is not permanently open, so differences in pressure can build up. In an airplane, for example, your ears may feel uncomfortable, and your hearing seems to fade. If you swallow or yawn, your Eustachian tubes open, your ears "pop" as the pressure equalizes, and your hearing returns to normal.

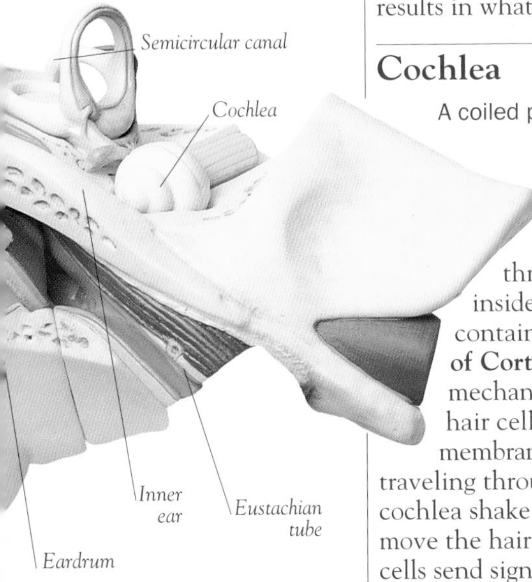

Middle and inner ear
The tiny ear ossicles carry vibrations from the eardrum to the inner ear.

Inner ear

A system of chambers containing sensory cells

The inner ear is also known as the **labyrinth**, because it is a mazelike collection of cavities and channels inside the skull. It is filled with fluid and consists of the cochlea, the semicircular canals ■, and a space called the vestibule ■. Vibrations from the eardrum are passed to the inner ear by the ear ossicles. One of these bones, the stapes, fits into a hole in the inner ear called the **oval window**. As the stapes moves in and out, it creates waves of pressure in the inner ear, and receptors convert these waves into nerve impulses ■. This results in what you hear as sound.

Cochlea

A coiled part of the inner ear

The cochlea is the part of the inner ear involved in hearing. There are three passageways inside it, one of which contains the spiral **organ of Corti**. This consists of mechanoreceptors ■ called hair cells attached to a membrane ■. Pressure waves traveling through the fluid in the cochlea shake the membrane and move the hair cells. The hair cells send signals to the brain, which interprets them as sounds.

Hair cell

A receptor that converts movement into nerve impulses

Hair cells are used in hearing, and also in the sense of balance ■. Each one has a bundle of tiny filaments. In the cochlea, these filaments are in contact with a membrane. When the membrane moves the filaments, the hair cells respond by triggering a nerve impulse. These impulses are carried to the brain by the **auditory nerve**, or **cochlear nerve**. The type of sound sensed by the brain depends on which hair cells are triggered.

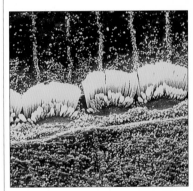

Hair cells
There are more than 15,000 hair cells in the organ of Corti. Each hair cell can contain up to 100 individual filaments.

Deafness

The inability to hear

The ear is a complicated organ, with many different parts. If just one part is not working normally, deafness can occur. In children, deafness is often caused by middle ear infections. Deafness in adults is often caused by earwax blocking the outer ear canal. In older people, deafness may be caused by changes in the moving parts of the ear. The eardrum becomes less flexible, and the ear ossicles are not so good at conducting movement to the inner ear. Deafness can also be caused by exposure to loud sounds.

Labels on *Middle and inner ear* diagram:
Ear ossicles, Eardrum, Semicircular canals, Oval window, Nerves to brain, Fluid in cochlea, Organ of Corti, Eustachian tube

Labels on lower diagram:
Semicircular canal, Cochlea, Inner ear, Eustachian tube, Eardrum

Touch & balance

You use the senses of touch and balance to detect pressure and movement. During a roller coaster ride, you feel the seat you are sitting on with your sense of touch, while balance tells you which way you are moving – even if your eyes are shut.

Touch

A sense that detects pressure

Touch is the sense that tells the body about its physical surroundings. It is a general sense ■, so it works through receptors ■ scattered all over the body. Some of these receptors respond to gentle touch, others detect firmer pressure.

Mechanoreceptor

A receptor that detects pressure

Mechanoreceptors produce nerve impulses ■ when they are pushed, pulled, or squeezed. They are used in touch, sensing posture, balance, and hearing ■.
Meissner's corpuscles near the skin's surface detect which part of the body is touching something.
Pacinian corpuscles deeper in the skin detect pressure. **Stretch receptors**, or **proprioceptors**, in muscles ■ and tendons ■ detect the body's posture. Hair cells ■ in the inner ear ■ are used in balance and hearing.

Mechanoreceptor in the skin
This Pacinian corpuscle, shown in cross section, detects pressure.

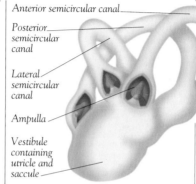

Anterior semicircular canal
Posterior semicircular canal
Lateral semicircular canal
Ampulla
Vestibule containing utricle and saccule

Balance organs in the inner ear
The semicircular canals and vestibule in the inner ear control your balance.

Balance

A sense that detects gravity and movement

Without the sense of balance, it would be impossible to stand up. The sense of balance detects the body's **equilibrium**, which is the way it is positioned relative to the pull of gravity, and the way it is moving. Both of these aspects are sensed in the inner ear.

Saccule

An organ that detects the pull of gravity

The saccule and its partner organ, the **utricle**, are chambers within part of the inner ear called the **vestibule**. They are lined with hair cells that are in contact with a layer of mineral crystals called **otoliths**. Gravity pulls on the crystals, and the hair cells sense their movement. The hair cells send signals to the brain, telling it which way is "up" and "down."

Semicircular canal

An organ that detects changes in the body's movement and position

The **anterior**, **posterior**, and **lateral semicircular canals** form part of the inner ear. These fluid-filled cavities are at right angles to each other. Each has a bulge at the base called an **ampulla**. This contains hair cells embedded in a jellylike knob called the **cupula**. When the head moves, the cupula and fluid in the canals also move. The hair cells sense this movement and send signals to the brain. If a person's speed keeps changing, the canals work nonstop, resulting in **travel sickness**, or **motion sickness**.

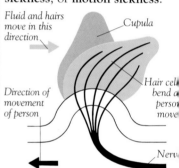

Fluid and hairs move in this direction
Cupula
Direction of movement of person
Hair cells bend as person moves
Nerve

Detecting changes in motion
The semicircular canals detect changes in speed, but not steady movement.

Pain

A warning sense that detects or warns of possible injury to the body

Pain is detected by nerve endings in the skin and other parts of the body. Sudden pain can warn you when you are in danger of injury. It may trigger a reflex ■ action that reduces the chance of harm.

See also

General sense 68
Hair cell 73 • Hearing 72
Inner ear 73 • Muscle 48
Nerve impulse 59 • Receptor 59
Reflex 63 • Tendon 49

Taste & smell

Taste and smell are two senses that go hand in hand. You use them to detect chemicals in the air and in food, and to distinguish things that are safe to eat from things that could be harmful.

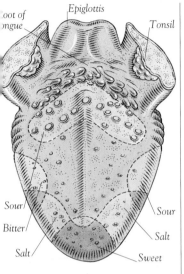

Taste areas of the tongue
The tongue is sensitive to only four tastes – sweet, sour, salt, and bitter.

Tongue

A muscular flap that moves food and is used in tasting and talking

The tongue is made of skeletal muscle. It is used to move and swallow ■ food, and to help the voice form sounds. The surface of the tongue is covered with small organs called taste buds, which are used to identify different foods.

Taste

A sense that detects dissolved chemicals

Taste is a sense that detects chemicals in food and drink. An unpleasant taste can act as a warning that a substance is not safe to eat. **Flavor** is a combination of taste and smell that enables you to identify many different foods.

Chemoreceptor

A receptor that detects chemicals

Chemoreceptors on your tongue and in your nose respond to chemicals in food or drink, and in the air. They contain microscopic hairs that trigger nerve impulses ■ when chemicals land on them. Some of these receptors respond to a wide range of chemicals, but others are much more specific.

Taste bud

A small organ that detects dissolved chemicals

Taste buds are small bundles of cells that contain clusters of chemoreceptors. Most taste buds are on the tongue, which has 10,000 in total. Each one has an opening, called a **taste pore**, that allows chemicals to enter the taste bud, so they can be tasted. Small organs called **papillae** project from the tongue. **Filiform papillae** are arranged in rows over the tongue's surface; **circumvallate papillae** and **fungiform papillae**, which both contain taste buds, are interspersed in between.

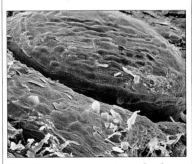

Papilla containing taste buds
Taste buds are located in the crevices at the edge of this circumvallate papilla.

Smell

A sense that detects airborne chemicals

Your sense of smell prepares your body to digest food by detecting when a meal is ready to eat. Smell works by sensing chemicals in the air, so it can only detect substances that give off airborne molecules. Smell signals are processed by the brain's limbic system ■, which also deals with emotions and memory.

Olfactory membrane

A thin, moist layer of cells used to detect airborne chemicals

Air normally enters the body through the nostrils and goes into the nasal cavity ■. Here, it flows past a lining called the olfactory membrane. This membrane contains chemoreceptors called **olfactory cells**. Cilia ■ on these cells collect airborne chemicals, and trigger impulses that are sent via the olfactory nerves ■ to the **olfactory bulbs**. These are paired swellings extending from the front of the brain. The brain interprets these signals as smells.

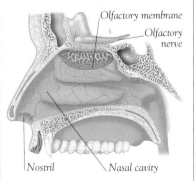

Olfactory membrane
Unlike taste buds, olfactory cells can detect many different chemicals.

See also

Cilia 27 • Limbic system 66
Nasal cavity 110 • Nerve impulse 59
Olfactory nerve 60 • Swallowing 120

Homeostasis

Whatever the conditions outside your body, conditions inside stay much the same. This stability is achieved through the processes of homeostasis, and is essential for life.

Homeostasis

The maintenance of stable conditions in a cell or in the body

Cells ■ work best in a very narrow range of conditions. They must not be too hot or too cold, and their surroundings must contain exactly the right balance of chemicals. Such conditions are maintained by homeostasis. Homeostasis means "staying the same." It involves chemical and physical processes that keep cells in a stable state, so that the body can function normally.

Environment

The surroundings in which a living thing exists

The body has two kinds of environment. The **external environment** is everything around the body, and is constantly changing. The **internal environment** is the environment around the body's cells. It is made up of blood ■, tissue fluid ■, and other liquids. Through homeostasis, conditions within the internal environment are kept stable, even when changes take place outside.

Keeping warm
Humans have learned to adapt to extreme conditions in their external environment.

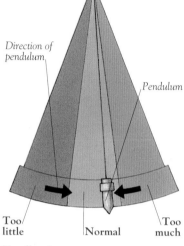

Direction of pendulum

Pendulum

Too little | Normal | Too much

Feedback system
Gravity brings a swinging pendulum back to its normal vertical position. In the same way, feedback systems return the body to its normal state.

Feedback system

A control mechanism that corrects unwanted changes

The body reacts to unwanted changes in its stable state by using control mechanisms called feedback systems. Each system monitors a particular factor, such as temperature, and keeps it within set limits. The body's feedback systems work with the help of hormones ■ and nerves ■. Nearly all of these systems involve **negative feedback**, which means that they reverse unwanted changes that occur. A few feedback systems, such as the one that controls contractions during labor ■, involve **positive feedback**, which means that they increase the change until a new state is brought about.

Thermoregulation

The control and adjustment of the temperature of the body

Humans are **endothermic** animals. This means that our bodies remain at a fairly constant temperature whatever the conditions outside. The average **body temperature** is 37°C (98.6°F), and it is controlled by a region of the brain ■ called the hypothalamus ■. The hypothalamus works like a thermostat. It makes the body generate more heat if it gets too cold, and lose more heat if it gets too hot. During a fever ■, the body's thermostat becomes "reset," so that it stays warmer than usual. This helps to combat the spread of microorganisms ■ during an infection ■. During **heatstroke** and **hypothermia**, the thermostat stops working, so the body becomes dangerously warm or cold. If this happens, outside help is often needed to bring body temperature back to normal.

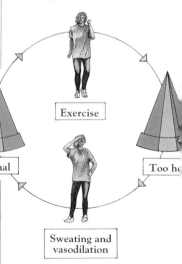

Exercise

Normal

Too hot

Sweating and vasodilation

Thermoregulation
Thermoregulation works by negative feedback. During strenuous exercise, the body's temperature rises. A feedback reaction causes perspiration and vasodilation, which cool down the body and return its temperature to normal.

Perspiration

The production of sweat

Perspiration, or **sweating**, is one of the body's ways of losing heat. When you perspire, sweat glands ■ secrete sweat onto your skin ■. The sweat evaporates, absorbing heat from the blood in your skin. Perspiration works best in dry air. If the air is very humid, sweat does not evaporate as easily, so it cannot remove as much heat.

Maintaining body temperature
After taking part in a marathon, this runner wraps himself in a silver thermal sheet. This reflects the body heat and reduces perspiration, which stops him from becoming too cold.

Vasodilation

The widening of blood vessels

If you exercise vigorously, your skin will quickly feel warm. This happens because the skin's blood vessels ■ widen or dilate, and carry more blood to the surface. The blood brings heat from the muscles. As it flows through the skin, some of its heat is lost to the outside. If you are cold, the opposite happens. The blood vessels get narrower, which decreases blood flow to the skin, and thereby reduces heat loss. This is called **vasoconstriction**.

Shivering

The repeated contraction of muscles to produce heat

Shivering is a sign that your body is trying to warm up. When you shiver, your muscles contract in repeated bursts and generate heat. This heat is then carried around the body in the blood. Shivering is often accompanied by goose pimples ■ on the skin.

Excretion

The process of eliminating waste

Like all living things, the body constantly generates chemical waste. This waste is potentially poisonous, and therefore has to be removed. It is formed in two ways. Some is generated by the body's metabolism ■, and some consists of substances that are taken in faster than they are used up. Waste is disposed of by the body's **excretory organs**. These include the lungs ■, which get rid of carbon dioxide, the skin, which gets rid of water and salts, and the liver ■, which gets rid of many different chemicals. The kidneys ■ get rid of water, salts, and **nitrogenous waste**, which contains nitrogen.

Respiratory control

The control of breathing rate and volume

All living cells need to be supplied with oxygen and to get rid of carbon dioxide. They do this with the help of blood. To carry out this task, blood must have the correct **oxygen/carbon dioxide balance**. This balance is a vital part of homeostasis. It is monitored by the respiratory center ■ in the brain. This area is very sensitive to the level of carbon dioxide in the blood.

Claude Bernard

French physiologist (1813–78)

Claude Bernard studied many of the processes that occur in the body, including digestion, respiration, and excretion. He discovered that cells need stable conditions, and that they work together to bring about this stability. Bernard's ideas were taken up by the American physiologist, **Walter Cannon** (1871–1945), who was the first person to use the word homeostasis.

Osmoregulation

The control and adjustment of osmotic pressure

Osmoregulation ensures that the osmotic pressure ■, or balance, of the body's fluids is kept at the right level, so that cells do not shrink or swell. It also ensures that different body parts contain the right amount of water. This is called **fluid balance**. The most important organs involved in osmoregulation are the kidneys. They control the osmotic pressure of blood, and this in turn controls the osmotic pressure of other body fluids, such as tissue fluid and cerebrospinal fluid ■.

See also

Blood 82 • Blood vessel 88 • Brain 64
Cell 26 • Cerebrospinal fluid 62
Fever 95 • Goose pimple 33
Hormone 78 • Hypothalamus 65
Infection 92 • Kidney 126 • Labor 140
Liver 122 • Lungs 112
Metabolism 102 • Microorganism 92
Nerve 58 • Osmotic pressure 29
Respiratory center 114 • Skin 32
Sweat gland 32 • Tissue fluid 96

Endocrine system

Two different systems keep the body working in a coordinated way. The nervous system works by using electrical impulses, while the endocrine system uses chemicals called hormones. Hormones usually work more slowly than nerves, but can have longer-lasting effects.

Endocrine system

A coordination system that releases chemical messengers into the blood

The endocrine system consists of nine major glands scattered throughout the body. Together, these glands produce dozens of chemical messengers, called hormones, which they release directly into the blood ■. The endocrine system plays an important part in homeostasis ■. Using chemical feedback ■, it regulates the metabolic and growth rates, sexual functioning, and development.

Hormone

A chemical messenger

Hormones are chemicals that the endocrine system uses to control body processes. Hormones travel around the body in the blood. When a hormone reaches its destination, called a **target cell**, or **target tissue**, it locks onto a specific site on the target's plasma membrane ■, called a **receptor site**. This causes chemical changes inside the cells, and triggers a specific action, such as the contraction of a smooth muscle ■. A hormone can have one target, or several.

Gland

A group of cells that produce and release chemicals

Glands make chemical substance and release them into or onto the body. This process is called **secretion**. The body has two different types of gland. **Exocrine glands** have ducts or channels, which secrete chemicals such as saliva or sweat into body spaces, or onto its surface. **Endocrine glands** do not have ducts. They secrete hormones into the blood

Endocrine gland
This type of gland secretes hormones directly into the blood.

Hormone secreted into bloodstream

Capillary

Exocrine gland
This type of gland secretes chemicals onto the body surfaces or into its spaces via ducts or channels.

Chemic secretic

Boo surfac

Chemic produce by gland

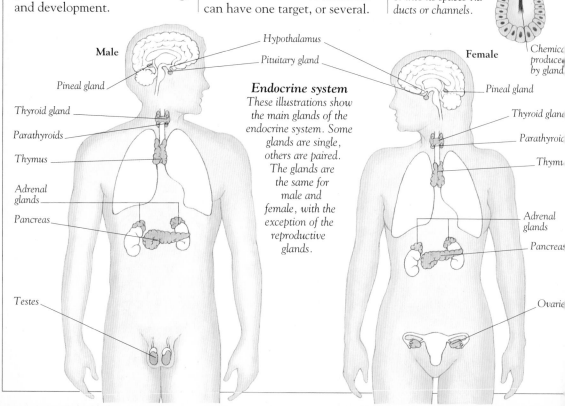

Male

Hypothalamus

Pituitary gland

Pineal gland

Thyroid gland

Parathyroids

Thymus

Adrenal glands

Pancreas

Testes

Endocrine system
These illustrations show the main glands of the endocrine system. Some glands are single, others are paired. The glands are the same for male and female, with the exception of the reproductive glands.

Female

Pineal gland

Thyroid gland

Parathyroid

Thymu

Adrenal glands

Pancreas

Ovarie

tuitary gland

endocrine gland at the base of
e brain

e pituitary gland is the
adquarters of the endocrine
tem. It releases at least nine
rmones that have important
ects on the body. Some of
ese control body functions
ectly; others trigger different
nds to produce hormones of
eir own. For example,
yroid-stimulating
rmone, also known as
yrotropin, or TSH,
mulates the thyroid gland ■.
drenocorticotrophic
rmone, or ACTH,
mulates the adrenal
nds ■. The pituitary
nd is divided into two
rts. The anterior lobe, or
enohypophysis, makes up
st of the gland, and releases
e majority of the hormones.
e smaller posterior lobe, or
urohypophysis, stores
rmones, but does not
ake them itself. It links
e endocrine system
th the nervous
stem ■ via the
pothalamus ■.

tuitary gland
e tiny, pea-sized
uitary gland is the most
portant endocrine gland.

Follicle-stimulating hormone

A pituitary hormone that promotes
the production of sex cells

Follicle-stimulating hormone, or
FSH, is produced by both men
and women. In men, FSH
stimulates the testes ■ to produce
sperm ■. In women, it stimulates
the ovaries ■ to produce fluid-
filled sacs called follicles ■,
which contain egg cells, or ova.

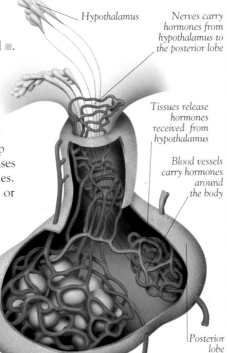

Hypothalamus

*Nerves carry
hormones from
hypothalamus to
the posterior lobe*

*Tissues release
hormones
received from
hypothalamus*

*Blood vessels
carry hormones
around
the body*

*Posterior
lobe*

Anterior lobe

Prolactin

A pituitary hormone that stimulates
the production of milk

Prolactin is one of several
hormones that stimulate milk
production, or lactation ■.
Breastfeeding stimulates the
pituitary gland to make more
prolactin, so that milk is made
for as long as the baby feeds.

Oxytocin

A pituitary hormone that stimulates
muscle contractions during birth

During birth ■, oxytocin
stimulates muscle contractions in
the uterus ■. These contractions
trigger the release of more
oxytocin. A positive feedback
reaction makes the cycle continue
until the baby is born. Oxytocin
also stimulates the breasts to
release milk when a baby feeds.

Antidiuretic hormone

A pituitary hormone that increases
the amount of water in the blood

A **diuretic** is a substance that
stimulates the body to produce
urine ■. Antidiuretic hormone,
also known as **vasopressin** or
ADH, has the opposite effect.
It increases the amount of water
that the kidneys return to the
blood, and makes arterioles ■
constrict. As a result, more fluid
is squeezed into a smaller space,
and the blood pressure ■ rises.

rowth hormone

pituitary hormone that
imulates growth

rowth hormone controls the
dy's growth by stimulating cell
vision ■. It also increases the
ood glucose level. If a child has
o little growth hormone, the
dy fails to grow normally. This
n result in a condition called
warfism. Too much growth
rmone can cause the body to
ow more than usual. The result
a condition called **gigantism**.

Luteinizing hormone

A pituitary hormone that helps
regulate the funtion of sex organs

Luteinizing hormone, or **LH**,
stimulates ovulation, which is
the release of egg cells by a
woman's ovaries. It also
stimulates an ovary to form a
corpus luteum ■. As the corpus
luteum develops, it releases sex
hormones ■ that prepare a
woman's body for pregnancy. In
men, LH triggers the testes to
produce male sex hormones.

See also

Adrenal gland 81 • Arteriole 88 • Birth 140
Blood 82 • Blood pressure 89
Cell division 30 • Corpus luteum 131
Feedback system 76 • Follicle 130
Homeostasis 76 • Hypothalamus 65
Lactation 141 • Nervous system 60
Ovary 129 • Plasma membrane 28
Sex hormone 81 • Sexual reproduction 128
Smooth muscle 49 Sperm 128 • Testis 128
Thyroid gland 80 • Urine 126 • Uterus 129

Continued over page ➤

Thyroid gland

A gland that controls metabolic rate and calcium levels in the body

The thyroid gland is located in the front of the neck, just below the larynx ■. It produces two hormones – **thyroxine** and **calcitonin**. Thyroxine contains iodine, and is the body's equivalent of an accelerator pedal. It speeds up the body's metabolic rate ■, makes its cells divide more rapidly, improves the performance of the nervous system ■, and increases heart rate and blood pressure. Calcitonin slows the rate at which bone is broken down. This decreases the amount of calcium that is dissolved in the blood.

Parathyroid gland

A gland that controls calcium levels in the blood

The parathyroids are four small paired glands surrounded by the thyroid gland. They produce **parathyroid hormone**, or **PTH**, which increases the rate at which bone is broken down. As a result, more calcium is released into the blood. Parathyroid hormone works in partnership with calcitonin from the thyroid gland. The two hormones have opposite effects. Through negative feedback ■, they keep the blood calcium level stable.

See also

Glucagon

A hormone that increases the level of sugar in the blood

Glucagon plays a vital part in maintaining the correct sugar level in the blood. It is made by the pancreas ■, a gland that forms part of the endocrine system ■ and the digestive system ■. The pancreas releases glucagon when the blood sugar level begins to fall. Glucagon makes cells release glucose, and helps convert glycogen ■, the form in which glucose is stored in the liver ■, back to glucose. As a result, the blood sugar level rises. Your blood contains enough glucose to keep you alive for just 15 minutes. However, as glucose is used up, more is released to take its place.

Color chart

Strip is dipped into urine sample

Urine testing
A urine test can be used to check for metabolic disorders within the body. For example, the presence of glucose in urine can be a sign of diabetes.

Color chart

Blood glucose test
A drop of blood is placed on a stick a held against a color chart. The result i used to determine the blood sugar leve A high reading can indicate diabetes.

Insulin

A hormone that reduc the blood sugar lev

Insulin is a protein ■ made by the pancrea It is released when the blood sugar level rises, ar reduces the sugar level in two ways. Firs it makes cells take up glucose Second, it make the liver store glucose by turning it int glycogen. Insulin and glucago have opposite effects. Together, they form a negative feedback system that keeps sugar levels within set limits. In people with diabetes ■, this control system does not work properly, and the may need daily injections of insulin to keep their blood glucose levels within safe limits

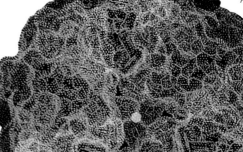

Molecule of insulin
This computer-generated image shows the structure of an insulin molecule. The small green circle at the center of the molecule represents a zinc atom. Crystals of insulin, consisting of 2 zinc atoms for every 6 insulin molecules, are stored in the pancreas.

◄ Continued from previous page

Sex hormone

A hormone that prepares the body for reproduction

Sex hormones are produced by the pituitary gland ■ and by the organs that contain sex cells ■. Several of these hormones are found in both sexes, but they have different effects in the male and female body. Unlike other hormones, sex hormones are not released until the age of about 10 or 11 years. During the period of growth called puberty ■, they stimulate a burst of development. This produces physical changes, called **secondary sexual characteristics**, which prepare the body for reproduction. In girls, these characteristics include enlargement of the breasts and widening of the pelvis ■. In boys, they include deepening of the voice and growth of facial hair.

Female sex hormone

A hormone that prepares the female body for reproduction

In women, sex hormones regulate ovulation ■ and equip the body for pregnancy. Many of them are released in a monthly cycle. Some female sex hormones are produced by the pituitary gland. Others, called **estrogens**, which produce secondary sexual characteristics, are made by the ovaries ■. **Progesterone**, which aids placenta function and helps prevent ovulation during pregnancy and lactation ■, is produced by the corpus luteum ■.

Male sex hormone

A hormone that prepares the male body for reproduction

The main male sex hormone is **testosterone**, which is released by the testes ■. This hormone produces secondary sexual characteristics and regulates the development of sperm ■.

Adrenal gland

A gland that increases metabolic rate and prepares the body for stress

There are two adrenal glands, or **suprarenal glands**, one on top of each kidney. These glands are divided into two parts, and each produces different hormones. The outer part, called the **cortex**, produces hormones called **corticosteroids**. These alter the concentration of ions ■ in the blood, and help to control metabolism ■. The inner part, called the **medulla**, produces the hormone epinephrine.

Epinephrine

A hormone that prepares the body for danger or stress

If you are suddenly alarmed or frightened, your heart pounds; your breathing becomes deep and rapid. These are just two effects of the hormone epinephrine, also called adrenaline. Epinephrine is very fast-acting, preparing the body for emergency action. It speeds up respiration ■ and heart rate ■, and diverts extra blood to the muscles. At the same time, it slows digestion ■, and makes the liver release glucose into the blood; thus, more fuel is available to make muscles contract. Epinephrine works with the nervous system to help you cope with danger or stress.

Jokichi Takamine

Japanese biochemist (1854–1922)

In 1901, Takamine became the first person to prepare crystals of a pure hormone. The hormone was epinephrine, which Takamine had taken from animal adrenal glands. At the time Takamine made his breakthrough, scientists were just beginning to understand how chemicals coordinate body processes. In 1902, the English physiologist **Ernest Starling** (1866–1927) isolated a hormone from the small intestine. In 1905, Starling coined the term hormone from a Greek word meaning "to stir up."

Roller coaster
On a thrilling roller coaster ride, the body's levels of epinephrine increase in response to the feelings of fear and excitement generated by the ride.

Blood

Blood is the body's liquid conveyor belt, and its most important line of defense. It flows past every living cell, delivering essential supplies and removing waste. An adult has up to 10 pints (6 liters) of this vital fluid.

Globin chain

Heme

Hemoglobin molecule
A hemoglobin molecule consists of a protein called globin arranged as four chains (purple and green). Each chain carries a pigment called heme (white). Heme groups combine with molecules of oxygen to form oxyhemoglobin.

Blood

A complex fluid that supplies and maintains the body's cells

Blood is a liquid connective tissue ■ that consists of billions of cells ■ suspended in a watery liquid. It keeps every part of the body supplied with oxygen and digested food, and it carries away waste products for excretion ■. Blood also carries heat around the body, and helps it to fend off attack by invading organisms.

Red blood cell

A blood cell that carries oxygen

There are millions of cells in a single drop of blood. Over 99 percent of these are red blood cells, or **erythrocytes**. A red blood cell is shaped like a doughnut, and has no nucleus ■. It contains hemoglobin, a protein that absorbs oxygen when the red blood cell passes through the lungs and releases it when the red blood cell passes through other parts of the body. Red blood cells are made in red bone marrow ■. They are smaller than most other cells in the body, but their shape gives them a relatively large surface area. This makes it easier for them to absorb oxygen and release it where it is needed. Each red blood cell lives for about four months, and is then replaced.

Hemoglobin

A protein that contains iron and carries oxygen

Hemoglobin is a transport protein ■ that contains iron. It is found in the cytoplasm ■ of red blood cells. In the lungs, molecules ■ of hemoglobin bind with oxygen to form a compound called **oxyhemoglobin**. Oxyhemoglobin carries oxygen around the body in the blood, and releases it wherever it is needed. Blood in arteries ■ contains lots of oxyhemoglobin, which makes it look bright red. Blood in veins ■ contains much less, and looks darker.

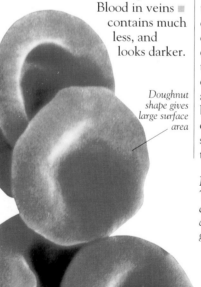

Doughnut shape gives large surface area

Anemia

A condition in which blood contains reduced levels of hemoglobin

Anemia is a common disorder that prevents blood from carrying its full amount of oxygen. An anemic person has either too few red blood cells, or too little hemoglobin in each one. As a result, people with anemia often look pale and easily become tired. Anemia is often caused by a shortage of iron, or a shortage of the vitamin needed to make red blood cells.

Red blood cells
This electron micrograph of red blood cells clearly shows their round, doughnutlike shape. Hemoglobin gives the cells their red color.

White blood cell

A blood cell that combats infection

White blood cells, or **leucocytes**, are larger than red blood cells, and much less numerous. They do not contain hemoglobin, but they do have a nucleus. Unlike red blood cells, white blood cells can change shape. They circulate in the blood, but may also squeeze through the walls of capillaries ■ to reach different parts of the body. There are many kinds of white blood cell, including lymphocytes, granulocytes, and monocytes. Their lifespans range from a few days to many weeks. Together, they make up a mobile defense force that protects the body against infection.

Lymphocyte
This electron micrograph shows a lymphocyte. Unlike the regularly shaped red blood cell, this white blood cell can change shape. This allows it to squeeze through the wall of a capillary and travel to different parts of the body.

Lymphocyte

A kind of white blood cell

Lymphocytes play a vital part in the body's defenses. Most kinds of lymphocyte produce proteins called antibodies ■, which prevent us from getting some diseases again, and help the immune system ■ to destroy bacteria ■ and other foreign cells. Lymphocytes develop in the lymphatic system ■, although they are originally formed in the bone marrow. They circulate in the blood, and other parts of the body.

Granulocyte

A kind of white blood cell with a lobed nucleus

When the body is invaded by bacteria, granulocytes are among the first defenders to arrive at the scene. Granulocytes develop in red bone marrow, and have nuclei with two or more large lobes. They are called granulocytes because their cytoplasm contains lots of small granules. Many granulocytes are **phagocytic**, which means that they engulf things by changing shape, and then digest them. In this way, they clear up foreign particles and bacteria, or dead body cells.

Monocyte

A kind of white blood cell with an unlobed nucleus

When the body becomes infected, monocytes are attracted to the site in large numbers. Here, they grow bigger and turn into cells called **wandering macrophages**. These engulf large quantities of bacteria and debris. Some parts of the body, including the spleen ■ and the lymph nodes ■, contain cells called **fixed macrophages**. These also engulf invading organisms, but remain in the same position.

See also

Antibody 98 • Artery 88 • Bacteria 92
Bone marrow 34 • Capillary 88 • Cell 26
Coagulation 84 • Connective tissue 19
Cytoplasm 26 • Excretion 77
Immune system 98 • Lymph node 97
Lymphatic system 96 • Molecule 20
Nucleus 27 • Osmotic pressure 29
Spleen 97 • Transport protein 24 • Vein 88

Platelet

A cell fragment that helps blood to form clots

Platelets are also called **thrombocytes**. They are about 30 percent of the size of red blood cells, and are usually round and flat with no nucleus. Platelets help blood to coagulate ■, and block small blood vessel wounds by sticking together to form a solid mass, called a **platelet plug**.

Blood plasma

The liquid part of blood

Plasma is a watery, yellowish liquid that contains dissolved substances, such as the products of digestion, ions, and **plasma proteins**. Plasma proteins include **fibrinogen**, which is essential for coagulation, and **albumin**, which helps to create most of the blood's osmotic pressure ■. Plasma also contains antibodies.

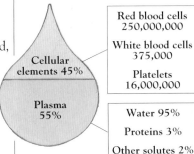

Red blood cells	250,000,000
White blood cells	375,000
Platelets	16,000,000

Cellular elements 45%

Plasma 55%

| Water 95% |
| Proteins 3% |
| Other solutes 2% |

A drop of blood
This diagram shows the approximate composition of a 0.003 inch³ (50 mm³) drop of blood.

How blood clots

A small skin wound, such as a cut, triggers a series of chemical reactions. A clot forms to seal the wound, and help limit the loss of blood and other fluids.

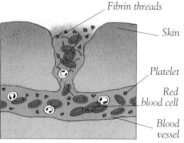

Fibrin threads
Skin
Platelet
Red blood cell
Blood vessel

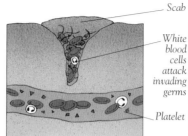

Scab
White blood cells attack invading germs
Platelet

1 *Platelets gather at the site of the wound and begin to stick together. Fibrin threads form a web and trap red blood cells.*

2 *The fibrin threads bind the red blood cells into a clot. The surface of the clot hardens into a scab, beneath which the wound heals.*

Coagulation

The process of clot formation

When a blood vessel ■ is cut, a solid lump called a **blood clot** forms at the site of the wound. The clot is formed when platelets ■ near the wound stick together to form a platelet plug ■. At the same time, substances released by the blood and damaged cells produce chemicals called **clotting factors**. These turn the soluble plasma protein ■ **fibrinogen** into the insoluble protein **fibrin**. Fibrin forms long strands that bind the blood cells and debris into a clot. A hard crust called a **scab** forms over the surface of the clot. People with the blood disorder **hemophilia** lack one clotting factor, which makes their blood slow to coagulate.

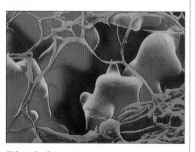

Blood clot
In this false-color electron micrograph, a red blood cell (orange) is enmeshed in a web of green fibrin threads.

Thrombosis

The formation of a clot inside a blood vessel

Blood does not normally clot inside blood vessels. But if a blood vessel becomes infected or clogged up by fat deposits, thrombosis may occur. The clot produced is called a **thrombus**. Thrombosis can be dangerous because the clot may be dislodged and carried around the body. If the clot becomes stuck in a blood vessel, it may block the blood supply to important organs. A **stroke** is a brain ■ disorder that is often caused by a thrombus. It usually affects one side of the brain, and prevents the part of the body it controls from working normally.

Anticoagulant

A substance that prevents the formation of blood clots

Clotting can be dangerous, so blood contains substances called anticoagulants that stop blood clots from developing when they are not wanted. These substances work by preventing fibrin from forming. People who are at risk from thrombosis, or who have had heart surgery, are often given extra anticoagulants to keep their blood in a fluid state.

Blood group

A class of blood that has characteristic proteins

Red blood cells ■ are covered with proteins ■ called antigens, while blood plasma ■ contains different proteins called antibodies ■. In humans, there are over 200 different blood group antigens, and they make up over 20 related blood groups, or **blood group systems**. The most important system, the **ABO system**, contains just two antigens, A and B, and four blood groups – A, B, AB, and O. People with type A blood have the A antigen, type B people have the B antigen, type AB people have both, and type O people have neither. People in the same blood group can exchange blood without any ill effects. But people from different blood groups often cannot exchange blood. Their blood may be **incompatible**, which means that the antigens and antibodies they contain will react with each other, and could stick together or agglutinate.

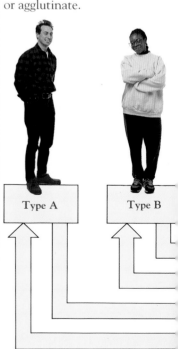

Type A

Type B

◄ Continued from previous page

Rhesus system

A blood group system first discovered in monkeys

The Rh (rhesus) factor system of blood groups was first discovered in rhesus monkeys. About 85 percent of the human population are Rh positive, which means that their blood cells carry the D antigen, or rhesus antigen. The remaining 15 percent do not have the anitgen and are Rh negative. Rhesus blood groups can cause problems during pregnancy. If a baby is rhesus-positive and its mother is rhesus-negative, the mother's antibodies may cross into the child's circulation ■ and attack its red blood cells. In severe cases, the baby has to be given a complete change of blood soon after it is born.

Giving and receiving blood

The types of blood each blood group can safely receive are shown below. Each group can receive blood from its own group. Types A and B can also receive from type O, while type AB can also receive from types A or B. Type O can give blood to all the other groups, but can receive it only from type O donors.

Type AB   Type O

Blood transfusion

The transfer of blood from one person to another

If the body loses a small amount of blood, it can soon make more to replace it. But if it loses a large amount, for example, more than 2.1 pints (1 liter), emergency action is needed to restore the volume of blood to its normal level. This is done by a transfusion of blood from another person, who is called a **blood donor**. The blood has to be examined to ensure that it does not contain bacteria ■ or viruses ■. Its blood group must also be known, so that it can be used without causing agglutination.

Blood transfusion
Blood from a donor is transfused into a vein in the arm. The patient's pulse, blood pressure, and temperature are monitored throughout the transfusion in case an adverse reaction occurs.

Agglutination

The clumping together of blood cells

In normal blood, red blood cells have a slippery surface and never stick together. But if blood from two people is mixed, the red cells may form clumps, or agglutinate. This happens because an immune response ■ takes place between proteins on the red blood cells and proteins in the plasma. Agglutination can be fatal, because the clumped cells can block blood vessels.

Karl Landsteiner

Austrian-American immunologist (1868–1943)

At one time, blood transfusions were very risky. In 1900, Landsteiner discovered that blood cells can clump together when blood is mixed, which explained why transfusions gave such unpredictable results. In 1909, Landsteiner worked out the ABO blood system, and, in 1940, he and other workers discovered the rhesus system. Landsteiner's work made transfusions safe, and has saved millions of lives.

Blood test

A test that analyzes the composition of blood

A blood test provides information about the state of a person's body, and can reveal details about their past medical history. By analyzing blood, a doctor can establish a patient's blood group, investigate their blood cells, and check what dissolved substances their blood contains. Unusual levels of some substances often show that a particular organ is not working normally. If the blood contains antibodies to a particular disease-causing organism, this shows that the body has been exposed to the disease at some time in the past.

See also

Antibody 98 • Bacteria 92
Blood plasma 83 • Blood vessel 88
Brain 64 • Circulatory system 90
Immune response 98 • Plasma protein 83
Platelet 83 • Platelet plug 83
Protein 24 • Red blood cell 82 • Virus 92

Heart

Throughout an average lifetime, the heart will beat more than two billion times, and will pump enough blood to fill over 100 full-sized swimming pools. Despite this amazing work load, a healthy heart never stops to rest.

Structure of the heart
The heart has four main chambers – the right and left atria and ventricles.

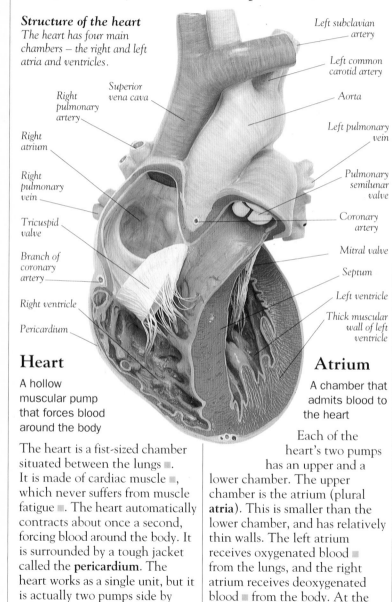

Left subclavian artery

Left common carotid artery

Superior vena cava

Right pulmonary artery

Right atrium

Aorta

Left pulmonary vein

Right pulmonary vein

Pulmonary semilunar valve

Tricuspid valve

Coronary artery

Branch of coronary artery

Mitral valve

Septum

Right ventricle

Left ventricle

Pericardium

Thick muscular wall of left ventricle

Heart

A hollow muscular pump that forces blood around the body

The heart is a fist-sized chamber situated between the lungs ■. It is made of cardiac muscle ■, which never suffers from muscle fatigue ■. The heart automatically contracts about once a second, forcing blood around the body. It is surrounded by a tough jacket called the **pericardium**. The heart works as a single unit, but it is actually two pumps side by side, separated by a muscular wall called a **septum**. One pump sends blood around the lungs; the other sends it around the body.

Atrium

A chamber that admits blood to the heart

Each of the heart's two pumps has an upper and a lower chamber. The upper chamber is the atrium (plural **atria**). This is smaller than the lower chamber, and has relatively thin walls. The left atrium receives oxygenated blood ■ from the lungs, and the right atrium receives deoxygenated blood ■ from the body. At the beginning of each heartbeat, the two atria contract, pushing the blood into the heart's two lower chambers, called the ventricles.

Ventricle

A chamber that expels blood from the heart

The ventricles are the hardest-working parts of the heart, and have thick muscular walls. They receive blood from the atria, and then pump it out of the heart on its journey around the circulatory system ■. The right ventricle pumps deoxygenated blood to the lungs, and the left one pumps oxygenated blood around the body. Because the oxygenated blood has farther to travel, the left ventricle is bigger and more powerful than the right.

Heart valve

A valve that allows blood to flow in one direction only

Heart valves stop blood from flowing backward. Without them, blood would try to flow in two directions at once. The heart has two sets of valves. Each valve has flaps that open to let blood flow forward, but automatically close to stop it from flowing back. The **tricuspid** and **mitral** valves prevent blood from flowing back from the ventricles to the atria. The **pulmonary semilunar** and **aortic semilunar** valves stop blood from flowing back to the ventricles after it has left the heart. If a heart valve fails to open and close properly, this reduces the flow of blood through the heart. Damaged valves can now be replaced by mechanical substitutes.

Open valve   Closed valve

How a semilunar valve works
The valve flaps open to allow blood through, but automatically closes afterward to prevent it from flowing back.

Heartbeat

A single cycle in the heart's pumping action

When your heart beats, it feels as though the whole of it contracts at once. But a complete heartbeat cycle has three stages that occur in a precisely timed sequence. During the first stage, called **diastole**, the atria fill with blood, and some blood flows through them into the ventricles. During the second stage, or **atrial systole**, the atria contract and empty their blood into the ventricles. A split-second later, during **ventricular systole**, the ventricles contract, and blood leaves the heart. Once this is complete, the heart enters diastole once more.

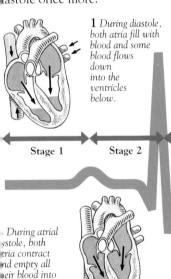

1 *During diastole, both atria fill with blood and some blood flows down into the ventricles below.*

3 *During ventricular systole, the ventricles contract and pump blood out of the heart and around the body. After this, the heart enters diastole once more.*

During atrial systole, both atria contract and empty all their blood into the ventricles.

Pulse

The momentary expansion of an artery following a heartbeat

When blood flows out of your heart, a wave of high pressure passes through your arteries ■. Your arteries expand slightly, and then return to their normal size. This momentary change in shape is what you feel as a pulse.

Electrocardiograph

An instrument that records the heart's electrical activity

An electrocardiograph is used to identify heart disorders. Electric terminals, or **electrodes**, are attached to the patient's chest and limbs. Pairs of electrodes are then connected in turn, so that each pair creates an electrical circuit that includes the body. The electrocardiograph records electrical changes in the circuit as the heart beats. This recording is called an **electrocardiogram**, or **ECG**.

Heartbeat cycle and ECG
The three stages in the heartbeat cycle correspond to different sections of the ECG graph.

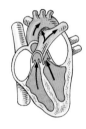

Stage 1 Stage 2 Stage 3

ECG graph

Heart rate

The number of heartbeats in a given time

Your heart has its own built-in rhythm, but it does not always beat at the same rate. Your heart rate, or **pulse rate**, varies according to how much oxygen you need. At rest, an adult's heart rate may be as low as 60 beats a minute. However, if you exercise vigorously, your heart rate can more than double. Heart rate is increased or decreased by the autonomic nervous system ■. It is also affected by hormones such as epinephrine ■.

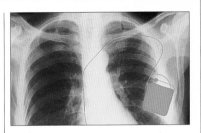

Artificial pacemaker
This X-ray shows an artificial pacemaker implanted in the chest.

Sinoatrial node

A small region of specialized muscle in the wall of the right atrium that triggers heartbeats

The sinoatrial node, also called the heart's **pacemaker**, produces electrical impulses. These spread through the heart and make the four chambers contract in the right sequence. Faulty pacemakers can now be replaced by electronic devices called **artificial pacemakers**.

Heart attack

Sudden death of part of the heart

During a heart attack, or **myocardial infarction**, part of the heart muscle dies because its oxygen supply is cut off. It is usually caused by a blockage of a coronary artery ■, often by a thrombosis ■. A severe heart attack may cause **fibrillation**, in which the ventricles contract quickly, disturbing the heart's rhythm. If this persists the person soon dies. After a heart attack, a person may suffer **heart failure**, in which the heart cannot pump enough blood around the body.

See also

Artery 88 • Autonomic nervous system 60
Cardiac muscle 48 • Circulatory system 90
Coronary artery 90 • Deoxygenated blood 88
Epinephrine 81 • Lung 112
Muscle fatigue 49 • Oxygenated blood 88
Thrombosis 84

Blood vessels

Like a system of pipes, blood vessels extend throughout the whole body, to keep cells alive. The largest blood vessels are thicker than a finger, but the smallest ones are so narrow that they can only be seen under a microscope.

Blood vessel

A tube that carries blood

Blood ■ is normally found only in blood vessels. The vessels form a network that passes close to all the body's cells. Blood circulates through this network, delivering nutrients and oxygen, and removing waste. The main types of vessel are arteries, veins, and capillaries. Each is shaped to work in a different way, but all have an inner tube or lumen ■, through which blood flows.

A network of blood vessels

Arteries divide into smaller vessels called arterioles, which themselves divide into capillaries. Blood then passes to venules and veins. Vein walls are thinner than artery walls, while capillary walls are thinner still.

Capillary network

Arteriole

Muscle fibers wrapped around arteriole

Artery

Very thin wall of capillary

Lumen

Inner layer

Middle layer of smooth muscle

Outer layer

Thick wall of artery

Venule

Vein

Inner layer

Middle layer of smooth muscle

Outer layer

Thin wall of vein

Lume

Capillary

An extremely fine blood vessel that supplies individual cells

Capillaries are the smallest bloc vessels. They are so narrow that red blood cells ■ often have to bend to squeeze through them. Capillaries collect oxygenated blood from arterioles, and carry it close to almost every cell in the body. Unlike other blood vessels, their walls are thin, just one cell thick. They allow dissolved substances to pass from the blood into tissue fluid ■, and back in the other direction. This blood then travels into venules and veins.

Artery

A blood vessel that carries blood away from the heart

Arteries have strong walls and carry blood under high pressure away from the heart. They divide into smaller vessels called **arterioles**. Arteries and some arterioles contain smooth muscle ■ in their walls. Arterioles have muscle fibers ■ wrapped around them. When the muscle contracts, the vessel narrows and allows less blood through. This change in shape, called vasoconstriction ■, helps to control body temperature. Most arteries and arterioles carry **oxygenated blood**, that is, blood rich in oxygen.

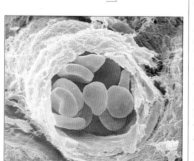

Red blood cells in an arteriole
This narrow arteriole allows only a few red blood cells to pass through at the same time; capillaries allow even fewer.

Vein

A blood vessel that carries blood toward the heart

After traveling through a capillary, blood passes into narrow blood vessels called **venules**, and then into veins. A vein has thinner walls than an artery. It usually contains **deoxygenated blood**, that is blood low in oxygen This gives veins a blue color. The blood in veins is under very low pressure, so the force driving it forward is weak. Many veins – particularly in the legs – contain **vein valves** to stop blood flowing back under its own weight.

Blood pressure

The pressure of blood in the circulatory system

Blood pressure is produced by the heart, and keeps blood moving. It is highest in arteries and lowest in veins. This difference ensures that blood flows in one direction only. Blood pressure rises each time the heart beats ■, then falls until the next beat begins. It also changes over longer periods of time, as the body adapts to different activities. It rises during excitement and exercise, and falls during rest. Blood pressure is controlled by the nervous system ■ and hormones ■.

Sphygmomanometer

An instrument that measures blood pressure with an inflatable cuff around the arm

Air is pumped into the cuff of a sphygmomanometer until the pressure stops blood flowing in the brachial artery ■. The air is then slowly released. When the cuff pressure falls just below that in the artery, blood starts to flow again. This can be heard through a stethoscope ■. The cuff pressure at this moment indicates the highest pressure in the artery. This is the **systolic blood pressure** – the pressure when the heart beats. When the pressure in the cuff is so low that it does not prevent flow in the artery at all, the lowest pressure can be measured. This is the **diastolic blood pressure** – the pressure between heartbeats.

Measuring blood pressure
To measure blood pressure, a cuff is placed around the arm and inflated.

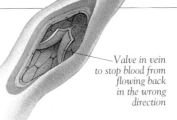

Valve in vein to stop blood from flowing back in the wrong direction

Hypertension

Abnormally high blood pressure

Hypertension is a common disorder of the circulatory system ■. Although it often produces no noticeable symptoms, it increases the risk of a heart attack ■ or a brain hemorrhage. Hypertension frequently has no obvious cause. It can often be treated by changes in diet ■ and by drugs that reduce the blood's volume.

Shock

A sudden reduction in blood flow

People involved in accidents often suffer from shock. This kind of shock is not a state of mind, but a failure of the body's blood supply. During shock, the blood pressure often drops to a low level, and a person may lose consciousness ■. Shock can be dangerous. Treatment includes keeping the body warm and, in severe cases where much blood has been lost, by giving blood transfusions ■.

Hemorrhage

The loss of blood from a blood vessel

Hemorrhage is a medical term that refers to any kind of bleeding. An **external hemorrhage** involves bleeding onto the body's surface, while an **internal hemorrhage** involves bleeding into the body's tissues. An internal hemorrhage can put pressure on nearby organs, and is particularly dangerous if it occurs in the brain ■.

Atherosclerosis

A disease that causes narrowing of the arteries

In atherosclerosis, layers of fatty material, called **plaques**, build up around artery walls. The flow of blood is reduced, and affected arteries become blocked by blood clots, which can cause a stroke ■ or heart attack. Atherosclerosis is linked to lifestyle. People who smoke, eat fatty food, and don't exercise much are more liable to get this disease.

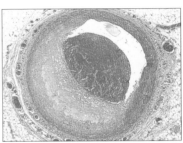

Plaque deposits in an artery
This artery has been narrowed by the plaque deposits around its walls. This restricts blood flow through the artery.

Varicose vein

A vein that has become misshapen

If a vein valve fails to close properly, a vein can fill up with blood and become stretched and twisted, or varicose. Varicose veins are a common disorder, particularly in pregnant women and old people. They are usually found in the legs, where gravity pulls blood the "wrong" way.

See also

Circulatory system

The circulatory system carries substances all over the body. Its network of blood vessels spreads from your bones and muscles to your teeth and toes. This system is powered by your heart.

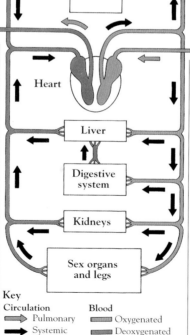

Coronary arteries
Branches of the two coronary arteries spread all over the surface of the heart.

Circulatory system

A system that maintains a constant flow of blood throughout the body

The circulatory system, or **cardiovascular system**, consists of the heart ■, the blood ■, and a vast network of blood vessels ■. It supplies nutrients and oxygen to all the living cells in the body and removes their waste. During its journey through the circulatory system, blood follows a figure-eight path. It flows first through the lungs and then around the body. Generally, a red blood cell ■ takes less than a minute to complete both parts of the circuit. Because the circulatory system has two loops, it is known as a **double circulation**.

Pulmonary circulation

The circulation from the heart around the lungs and back again

In the pulmonary circulation, deoxygenated blood ■ travels to the lungs ■ via two arteries ■ called the **right pulmonary artery** and the **left pulmonary artery**. As it passes through capillaries ■ in the lungs, it collects oxygen and releases carbon dioxide. The oxygenated blood ■ then returns to the heart through the two **left pulmonary veins** and the two **right pulmonary veins**, so that it can then be pumped around the body. The pulmonary circulation is unusual because arteries normally carry oxygenated blood and veins ■ normally carry deoxygenated blood, but here it is the opposite way around.

Systemic circulation

The circulation from the heart around the body and back again

The systemic circulation carries oxygenated blood to all parts of the body except the lungs, and returns deoxygenated blood to the heart. Blood leaves the heart through the aorta, and returns through two veins called the venae cavae.

Head, neck, and arms

Lungs

Heart

Liver

Digestive system

Kidneys

Sex organs and legs

Key
Circulation
⟹ Pulmonary
⟹ Systemic

Blood
▭ Oxygenated
▬ Deoxygenated

Circulatory routes
The pulmonary and systemic circulations pump blood around the whole body.

Coronary artery

An artery that supplies the heart

Although the heart contains blood, it needs its own blood supply to provide it with oxygen. Two coronary arteries branch off the aorta and fan out over the heart. Once the blood has flowed through the heart tissue, it is collected by a vein called the **coronary sinus**. This empties blood into the right atrium ■. The **coronary circulation** keeps the heart working. If a coronary artery is blocked by a blood clot, or **coronary thrombosis**, the result may be a heart attack ■.

Aorta

The main artery through which blood leaves the heart

The aorta is the largest artery in the body. In an adult, it is about 1 inch (2.5 cm) across at its widest point. It receives blood under high pressure from the left ventricle ■ of the heart. The aorta has tough but elastic walls to prevent it from bursting. After leaving the heart, the aorta travels up the body, before turning through a 180° bend and traveling down in front of the backbone toward the legs. The arch of the aorta connects with arteries that supply the head and arms; the lower part connects with arteries that supply the trunk and legs.

Vena cava

One of a pair of veins that empty blood into the heart

Blood returns from the body to the heart through two large veins that can measure up to 1 inch (2.5 cm) across. The **superior vena cava** collects blood from the head, neck, arms, and upper part of the trunk, while the **inferior vena cava** collects blood from the rest of the body. Both of these veins empty blood into the right atrium of the heart. The name vena cava means "hollow vein."

Carotid artery

One of the four main arteries that supply the head

Most arteries lie deep in the body, where muscles and bones protect them from injury. But the four carotid arteries are near the surface of the neck, next to the trachea ■. They supply oxygen to the brain. Injury to the carotid arteries can make a person lose consciousness ■ through reduced blood supply to the brain.

Jugular vein

One of the three main veins that drain the head

Jugular veins carry blood from the head back toward the heart. Jugular veins are not as vulnerable as carotid arteries because they carry blood under lower pressure.

Femoral artery

An artery that supplies the leg

After passing through the diaphragm ■, the aorta divides into two **iliac arteries**. Each of these becomes a femoral artery, which travels down the leg close to the thighbone, or femur ■. The femoral artery divides into a number of smaller arteries, and these supply oxygen to the leg.

Femoral vein

A vein that drains the leg

The femoral vein is a large vein in the upper leg. It collects blood from smaller leg and foot veins, and passes it to the **iliac veins** in the groin, and then to the inferior vena cava. Another leg vein, the **great saphenous vein**, is the longest vein in the body, and runs from the groin to the feet.

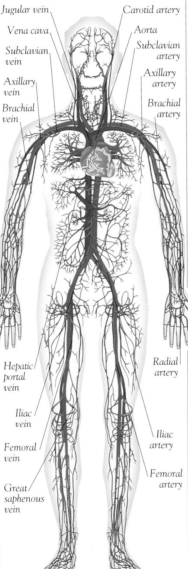

The body's main blood vessels
The main arteries and veins branch out to form a vast network of blood vessels.

Labels on figure: Jugular vein, Carotid artery, Vena cava, Aorta, Subclavian vein, Subclavian artery, Axillary vein, Axillary artery, Brachial vein, Brachial artery, Hepatic portal vein, Radial artery, Iliac vein, Iliac artery, Femoral vein, Femoral artery, Great saphenous vein

Brachial artery

An artery that supplies the arm

The names of blood vessels often change as they run through the body. The main artery supplying the arm starts as the **subclavian artery**, under the collarbone, or clavicle ■. It becomes the **axillary artery** as it travels through the armpit, or axilla ■, before it turns into the brachial artery, which runs down the arm itself. The brachial artery has many branches, including the **radial artery**, which lies next to the radius ■. This artery is close to the surface of the wrist, and this is where you can normally feel a pulse ■.

Brachial vein

A vein that drains the arm

The brachial vein is one of the large veins in the upper arm. It receives blood from the smaller veins in the hand and forearm, and passes it to the **axillary vein** near the armpit, the **subclavian vein** near the clavicle, and then to the superior vena cava.

Hepatic artery

The artery that supplies the liver

The hepatic artery supplies the liver ■ with oxygen-rich blood from the heart, while the **hepatic portal vein** supplies the liver with nutrient-rich blood from the digestive system ■. Together, these make up the **hepatic circulation**.

See also

Artery 88 • Atrium 86 • Axilla 14
Blood 82 • Blood vessel 88 • Capillary 88
Clavicle 40 • Consciousness 66
Deoxygenated blood 88 • Diaphragm 114
Digestive system 116 • Femur 41
Heart 86 • Heart attack 87 • Liver 122
Lung 112 • Oxygenated blood 88
Pulse 87 • Radius 41 • Red blood cell 82
Trachea 111 • Vein 88 • Ventricle 86

Health & disease

The human body is like a complex piece of machinery. Most of the time it runs smoothly, but sometimes it becomes damaged and breaks down. This damage can be caused by problems inside the body, or by invaders from outside.

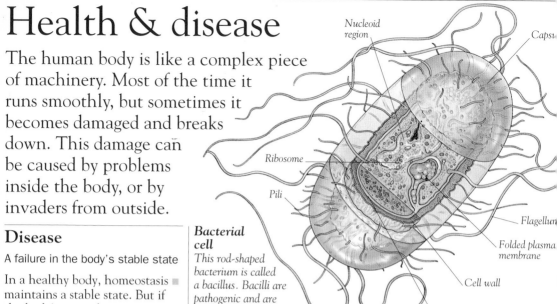

Nucleoid region

Capsu

Ribosome

Pili

Flagellum

Folded plasma membrane

Cell wall

Plasma membrane

Disease

A failure in the body's stable state

In a healthy body, homeostasis ■ maintains a stable state. But if the body's control systems stop functioning normally, homeostasis breaks down. This change in the body's state is called a disease. Some diseases are short-lived, and the body quickly overcomes them. Others are more serious, and outside help may be needed to fight them or to reduce their effects.

Pathogen

A disease-producing organism

A pathogen is any organism that can invade the body and cause disease. Pathogens attack the body's cells, or release powerful poisons called **toxins**. Nearly all pathogens are microscopic living things, called **microorganisms**. The most important pathogens are bacteria and viruses, although some diseases are caused by microscopic fungi or by single-celled organisms called **protozoa**.

See also

Alimentary canal 116 • Cancer 31
Cold 144 • Diabetes 146 • Gene 132
Homeostasis 76 • Influenza 144
Measles 145 • Nucleic acid 25
Protein 24 • Skin 32 • Syphilis 145

Bacterial cell

This rod-shaped bacterium is called a bacillus. Bacilli are pathogenic and are responsible for diseases such as typhoid and tetanus.

Bacteria

Microscopic, single-celled organisms

A healthy body contains at least 100,000 billion bacteria (singular **bacterium**), which are commonly known as **germs**. Germs are normally found on body surfaces, such as the skin ■, and in the alimentary canal ■. They usually do no harm. But bacteria that enter the body can become **pathogenic,** which means they cause disease.

Virus

A package of chemicals that invades a living cell

Viruses are much smaller and simpler than bacteria. Each one consists of a small amount of nucleic acid ■, surrounded by a protein ■ coat. When a virus infects a cell, its nucleic acid takes over the cell's chemical processes. The cell is forced to make copies of the virus, and is often destroyed in the process. Viruses do not show any signs of life outside cells.

Infection

The growth of disease-causing organisms in or on the body

The human body is an ideal environment for microorganisms. If a microorganism manages to break through the body's defenses, it may grow and reproduce quickly inside the body. The result is an infection. Some untreated infections develop into an infectious disease, which can spread to other people.

Incubation period

The interval between infection and the start of a disease

After a pathogen successfully invades the body, some time passes before a disease develops. This interval is known as an incubation period. Some pathogens have an incubation period of a few hours or days, but others have an incubation period of between one and a number of years. The pathogen may be passed on to other people during this time.

fectious disease

disease caused by pathogens

n infectious disease is one that
ou can catch by coming into
ntact with pathogens. Infectious
eases are spread in several
ys. Some are spread by direct
ysical contact. Many are spread
droplet transmission, which
ans they are carried in small
oplets of mucus when someone
eezes. A limited number are
read by **vectors**, which are
imals such as mosquitoes.

reading disease
osquitoes can spread disease when they
e their victims to feed on their blood.

ommunicable disease

infectious disease that can be
read from one person to another

ommunicable diseases are
ssed on when a pathogen
reads from one person to
other. This can happen either
rectly, or through an animal
ctor. Most infectious diseases
l into this category. A **non-**
mmunicable disease cannot be
read directly from person to
rson. It is often caused by
icroorganisms that do not
rmally live in or on the body.

ontagious disease

infectious disease that is easily
read from one person to another

ontagious diseases, such as
fluenza ▪, measles ▪, and
lds ▪, are easily passed on from
rson to person. They often
cur in outbreaks, or **epidemics**,
hich affect many people in the
me place at the same time.

Noninfectious disease

A disease that is not caused
by pathogens

A noninfectious disease is one
that you cannot catch by being
exposed to pathogens. Instead, it
is triggered by characteristics
coded into your genes ▪, or by
nonliving factors in your
environment, such as hazardous
chemicals. Noninfectious
diseases include diabetes ▪ and
most kinds of cancer ▪.

Symptom

An indication of a disease

Diseases affect the body in
different ways, and each one
produces its own set of
symptoms. For example, a sore
throat and runny nose are
common symptoms of a cold,
while headaches and aching
joints are common symptoms of
influenza. Symptoms are usually
noticed by the sufferer of a
disease. A **sign** is an indication
of a disease that is observed by a
doctor, rather than by a patient.

Sneezing
Sneezing is a symptom of a cold. Colds
can be transmitted in droplets of mucus.

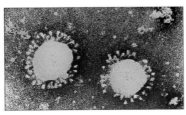

Cold virus
The virus above causes the common
cold. It is spread by droplet transmission.

Paul Ehrlich

German medical
scientist
(1854–1915)

As a medical
student, Paul
Ehrlich became interested in
newly discovered chemical
dyes, which stained some
kinds of cells but not others.
Ehrlich wondered if similar
chemicals could be used as
"magic bullets," destroying
invading bacteria but leaving
the body's cells unharmed. For
five years, he tested hundreds
of chemicals without success.
But in 1909, he discovered
salvarsan, a substance that
killed the bacteria responsible
for a disease called syphilis ▪.
This substance was the first
synthetic drug to be used.

Diagnosis

The identification of a disease or
disorder by a medical practitioner

To make a diagnosis, a doctor
examines a patient for symptoms
and signs, and checks his or her
medical history. The doctor will
then decide on the appropriate
treatment, or **therapy**. The
doctor's assessment of how the
patient will fare is a **prognosis**.

Making a diagnosis
A physical examination often helps a
doctor to identify a patient's illness.
Further tests may confirm the diagnosis.

Disease defenses

From the moment that you are born, your body is under attack from microscopic invaders. Fortunately, it is equipped with a wide array of defenses. Some of these defenses block all kinds of invaders, while others lock on to particular targets and destroy them.

Resistance

The ability to ward off a disease

Resistance keeps the body free from infection ▪. There are two main kinds . **Specific resistance** consists of a series of defenses that identify particular pathogens ▪, and then disable or destroy them. This kind of resistance is provided by the immune system ▪. **Nonspecific resistance** consists of general mechanical or chemical barriers that prevent invaders from entering and infecting the body.

Mechanical resistance

Nonspecific resistance through physical means

Skin ▪ is the body's most important means of mechanical resistance. Its outermost layer is made up of dead cells ▪, and these form a barrier that most microorganisms ▪ cannot cross. Mechanical defenses also protect the body's other exposed surfaces. A steady flow of tears washes away pathogens from the eyes. Cells lining the trachea ▪ trap pathogens in sticky mucus ▪, and then sweep them away with tiny hairs called cilia ▪.

Chemical resistance

Nonspecific resistance through chemical means

The body produces a range of toxic chemicals to protect it from invading bacteria ▪. For example, skin is chemically protected by an oily, acidic substance called sebum ▪. Sebum kills many bacteria, and prevents others from multiplying. Sweat ▪, tears, and saliva contain a chemical called **lysozyme**, which kills some kinds of bacteria by breaking down their cell walls. Another kind of chemical resistance occurs in the stomach ▪, which secretes hydrochloric acid. This kills many kinds of bacteria on contact.

Sweating
Sweat is a form of chemical resistance.

Interferon

A protein that prevents viruses from copying themselves

Interferons are proteins ▪ that provide the body with non-specific resistance against many kinds of viral infection. They are produced by infected cells, and stimulate neighboring cells to produce enzymes ▪ that prevent the virus ▪ from copying itself. The virus gradually dies out.

Inflammation

A defensive response to infection

Inflammation occurs when damaged body cells trigger the release substances such as **histamines** into the blood. These substances cause widening of the blood vessels, or vasodilation ▪. The increased blood flow to the inflamed area makes it red and swollen. Phagocytes are attracted to the infection site. They destroy the invaders, and the inflamed area returns to normal.

Inflammatory response
Inflammation is the body's response to infection. The infected area often feels warmer than usual, and may hurt.

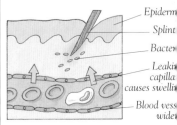

Epiderm
Splint
Bacter
Leaki
capilla
causes swelli
Blood vess
wide

1 *Damaged cells and the appearance of bacteria near a wound trigger the release of substances called histamines.*

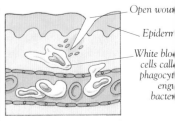

Open woun
Epiderm
White blo
cells call
phagocyt
engu
bacter

2 *The number of white blood cells at the infection increases to engulf the invaders.*

Phagocyte

cell that ingests cells and debris

Most phagocytes are white blood cells ▪. They enable the body to dispose of unwanted matter, such as pathogens, dirt that enters the body in cuts, and dead body cells. Phagocytes engulf the matter and then digest it in a process called phagocytosis ▪. When active phagocytes die, they sometimes form a whitish fluid called **pus**.

Complement system

group of blood proteins that help disable invading organisms

The complement system is made up of more than 24 proteins that circulate in the blood. Complement proteins are non-specific, so they are activated by a wide range of pathogens. They cause inflammation, and alert phagocytes. The complement system also moves into action when the immune system repels attackers. The proteins break open the cell membranes of the invaders, making them burst.

Bacterial flora

The harmless bacteria normally found on the body's surfaces

Many parts of the body, such as the skin, mouth, and intestines, harbor vast numbers of bacteria. These protect the body against other microorganisms, and help keep it healthy. But if any of these bacteria invade the body's tissues, they can cause disease.

Skin bacteria
This electron micrograph shows a colony of Staphylococcus aureus, a bacterium commonly found on the skin.

Fever

A state of abnormally high body temperature

A fever is caused when white blood cells release proteins called **pyrogens**, which reset the body's "thermostat" in the hypothalamus ▪. This makes the body temperature rise to as much as 104°F (40°C). A high body temperature makes it difficult for bacteria to multiply, but it can also cause a state called **delirium**, in which the brain begins to work abnormally.

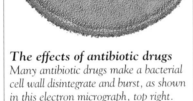

The effects of antibiotic drugs
Many antibiotic drugs make a bacterial cell wall disintegrate and burst, as shown in this electron micrograph, top right.

Drug

A chemical that affects the functions of the body, or the progress of a disease

Drugs are the most important weapons in helping the body to fight disease. There are many different kinds. **Antibiotics** are substances that kill bacteria, or prevent bacteria from infecting the body. **Analgesics** are drugs used to relieve pain ▪. Some act against mild pain such as a headache or toothache, but others, called **narcotics**, tackle severe pain. Many drugs are potentially poisonous, and have to be taken in controlled doses.

See also

Bacteria 92 • Cell 26 • Cilia 27
Enzyme 24 • Hypothalamus 65
Immune system 98 • Infection 92
Microorganism 92 • Mucus 19
Pain 74 • Pathogen 92 • Phagocytosis 29
Protein 24 • Sebum 33 • Skin 32
Stomach 121 • Sweat 33 • Trachea 111
Vasodilation 77• Virus 92
White blood cell 83

Antiseptic

A chemical that disinfects the skin

When bacteria enter a wound and begin to multiply, the wound may become **septic**, which means that cells around it begin to die and break down. Antiseptics kill bacteria on contact, before they can cause infection. They are used to disinfect deep cuts, and also to kill bacteria on the skin before an operation.

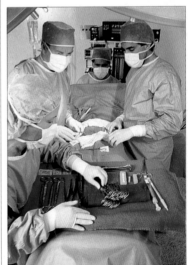

Before an operation
The skin must be thoroughly cleaned with antiseptic before a surgical cut can be made, in order to ensure that it is free of bacteria. Surgical instruments must also be disinfected to avoid the risk of infection.

Lymphatic system

Fluid constantly leaves your bloodstream, and flows through the spaces that surround your cells. Your lymphatic system drains some of this fluid, and filters out anything that might cause disease.

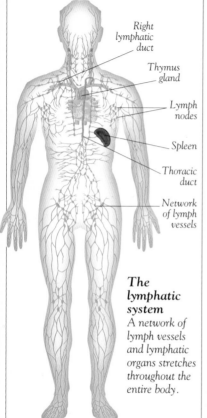

Right lymphatic duct

Thymus gland

Lymph nodes

Spleen

Thoracic duct

Network of lymph vessels

The lymphatic system
A network of lymph vessels and lymphatic organs stretches throughout the entire body.

Lymphatic system

A one-way system of channels that drains fluid and fights infection

The lymphatic system consists mainly of a network of tubes. These reach all over the body and drain excess fluid from the spaces between cells ■. The fluid flows into two ducts that empty it into the bloodstream. The lymphatic system is one-way; it returns fluid to the bloodstream, but it cannot collect it from there. It also helps the immune system ■ to fight infection ■, and carries digested fats ■ away from the intestines ■.

Extracellular fluid

Any fluid found outside cells

The most important types of extracellular fluid are tissue fluid, lymph, and blood plasma ■. Other kinds include cerebrospinal fluid ■, synovial fluid ■, and mucus ■. Together, these make up 30 percent of the body's fluid. The remaining 70 percent is made up of fluid contained in cells, called **intracellular fluid**.

Tissue fluid

A fluid that bathes living cells

Tissue fluid is also known as **intercellular fluid** or **interstitial fluid**, as it is found in the small spaces, called **interstices**, between cells. It is formed by liquid that seeps out of blood capillaries ■. Most of this liquid returns to the capillaries, but the rest enters the lymphatic system. Tissue fluid is similar to blood plasma, but contains less protein ■. It transfers substances between blood and cells, and creates the environment that cells need to survive.

Lymph

The fluid that flows through the lymphatic system

When tissue fluid enters the lymphatic system, it is called lymph. Lymph contains many dissolved substances, white blood cells ■ called lymphocytes ■, and the remains of microorganisms ■ that invade the body. Lymph often has a milky color because it contains microscopic globules of fat, particularly after a meal.

Lymph vessel

A tube that carries lymph

Lymph vessels have "leaky" or permeable walls, so that they ca drain fluid from the surrounding tissue cells. The fluid is collecte by small vessels called **lymph capillaries**, or lacteals, in the small intestine ■. The fluid is passed on to larger vessels called **lymph trunks**, and finally to the two largest vessels called **lymph ducts**: the **thoracic duct** and the **right lymphatic duct**. Most lymph flows through the thoracic duct, which empties into a vein near the heart. The lymphatic system, unlike the circulatory system ■, has no pump. Lymph pushed through the vessels whe nearby muscles contract. Valve in the vessels stop the fluid from flowing backward.

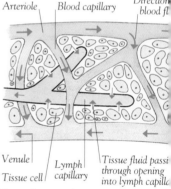

Arteriole | *Blood capillary* | *Direction blood fl*

Venule | *Lymph capillary* | *Tissue fluid passi through opening into lymph capilla*

Tissue cell

Lymph vessels
Lymph capillaries have small openings that allow them to collect excess tissue fluid around the body's cells.

Lacteal

A lymph vessel in the small intestine

Lacteals are lymph vessels that lie inside projections called villi ■ on the wall of the small intestine. Lacteals collect tiny globules of fat from the small intestine. The fat then travels through the lymphatic system, and is slowly emptied into the bloodstream.

ymphatic organ

1 organ that forms part of the mphatic system

s well as the network of lymph essels, the lymphatic system also ontains a number of organs. The mph nodes screen out unwanted bstances that flow through the 'stem, while the spleen and 1ymus help to produce cells that ght infection. Bone marrow ■ so plays a part in the lymphatic 'stem by forming lymphocytes.

ymph node

bean-shaped swelling in a mph vessel

ymph nodes work like filters. ach one is surrounded by a ough capsule, and contains a ense network of fibers. Lymph ows into the node, and foreign ells and debris are screened out nd destroyed. This cleaning-up rocess is carried out both by the bers, and by special white blood ells, including lymphocytes and hagocytes ■. Once the lymph as been filtered, it flows out of ne node. Lymph nodes occur hroughout the lymphatic system, ut they are numerous in the rmpits and the groin. During an nfection, they an swell up nd become tender a condition called swollen glands."

Spleen

An organ that fights infection and removes worn out red blood cells

The spleen is an oval organ beside the stomach. It forms part of the lymphatic system, and produces lymphocytes and phagocytes to fight infections. Before birth, the spleen also produces red blood cells ■. After birth, the spleen usually stops making new red blood cells, and instead filters the blood to remove old cells that are no longer working effectively.

Thymus gland

An organ that produces hormones and cells to fight infection

The thymus is a gland ■ beneath the upper part of the breastbone, or sternum ■. It is very large in young children, but by adulthood it is comparatively smaller. The thymus gland is involved in two different body systems. As part of the lymphatic system, it helps to produce T cells ■, which fight infections. As part of the endocrine system ■, it produces hormones that encourage T cells to mature.

Tonsil

A mass of lymphatic tissue at the back of the mouth

There are two tonsils at the back of the roof of the mouth, and a smaller pair at the base of the tongue ■. They help to protect the throat and airways from infection. The tonsils grow to full size during childhood, when they can often become inflamed.

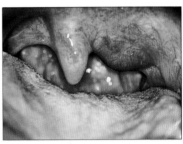

Tonsil in the throat
The tonsil is the flap of tissue hanging down at the back of the throat.

Adenoid

A mass of lymphatic tissue on either side of the nasal cavity

The two adenoids protect the upper part of the respiratory tract ■. If they become swollen, they can narrow the connection between the nose and throat. This can affect someone's voice, giving them a nasal or "adenoidal" sound.

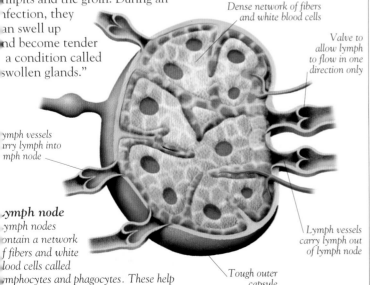

Dense network of fibers and white blood cells

Valve to allow lymph to flow in one direction only

ymph vessels rry lymph into mph node

ymph node
ymph nodes ontain a network f fibers and white lood cells called ymphocytes and phagocytes. These help o screen out debris and foreign material.

Lymph vessels carry lymph out of lymph node

Tough outer capsule

See also

Blood plasma 83 • Bone marrow 34
Capillary 88 • Cell 26
Cerebrospinal fluid 62
Circulatory system 90
Endocrine system 78 • Fat 23
Gland 78 • Immune system 98
Infection 92 • Intestine 124
Lymphocyte 83 • Microorganism 92
Mucus 19 • Phagocyte 95 • Protein 24
Red blood cell 82 • Respiratory tract 110
Small intestine 124 • Sternum 37
Synovial fluid 44 • T cell 99 • Tongue 75
Vein 88 • Villus 124
White blood cell 83

Immune system

The immune system is a collection of special cells that protects the body from invading organisms. It keeps information on all intruders, and, should they reappear, reacts with lightning speed to destroy them.

Fighting disease
When disease-causing organisms enter the body, defense systems begin to work.

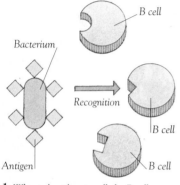

1 *When a lymphocyte called a B cell recognizes a specific antigen on the surface of a bacterium, it begins to multiply rapidly.*

Immune system

A system that produces specific defenses against disease-causing organisms

The immune system provides the body with specific resistance ■, by defending it against particular invading organisms. It uses white blood cells called lymphocytes ■ to recognize and attack foreign substances on invading cells. The invader is usually destroyed. The immune system also memorizes the pathogen ■ that causes the disease. In this way, you become **immune** to that disease, which means that you cannot suffer from it again. Humans have **natural immunity** to some diseases at birth, and develop **adaptive immunity** to others throughout life. The adaptive immune system has two parts – humoral immunity and cellular immunity.

Antigen

A foreign substance in the body

An antigen is any foreign substance that can trigger the immune system into action. Most antigens are proteins ■, and they are usually found on the surface of pathogens, such as bacteria ■ or viruses ■.

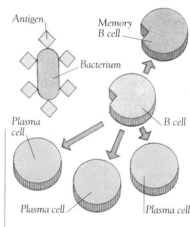

2 *The activated B cells turn themselves into plasma cells and memory B cells.*

Antibody

A protein that attaches itself to a specific foreign substance

Antibodies are also called **immunoglobulins**. They are made by lymphocytes, and are found in blood and other body fluids. Each type has its own chemical pattern, and is able to lock onto a particular antigen. In this way, it helps to disable a pathogen, or mark it so that it can be destroyed by phagocytes ■ or by the complement system ■.

Immune response

The response of the immune system to a foreign substance

When the immune system meets an antigen for the first time, it produces antibodies against it. This is the **primary response**. It can take a few days to occur, in which time the disease caused by the antigen may develop. If the immune system encounters the same antigen again, however, the correct antibodies are produced very quickly, and in much larger amounts. This is the **secondary response**. It is usually fast enough to overpower an invader so the body does not suffer from the same disease again.

Humoral immunity

A form of immunity that works by using antibodies

Humoral immunity is the part of the immune system that uses antibodies to destroy invaders. It is also known as **antibody-mediated immunity**. This form of defense works against most of the bacteria that infect the body, and also against some viruses.

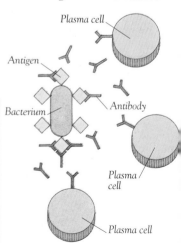

3 *The plasma cells release antibodies that travel in the blood to the site of infection. The antibodies lock onto the antigens on the bacterium, and disable the bacterium, or mark it for destruction.*

cell

cell used to produce antibodies

cells are lymphocytes that are
volved in humoral immunity.
When an invader appears, some
cells turn into **plasma cells**,
which are cells that produce the
orrect antibodies to attack the
vader. Once this has occurred,
her B cells turn into **memory**
cells. These "memorize" the
vader's antigens, so that
ntibodies can be made quickly if
reappears. Plasma cells live for
st a few days, but memory B cells
ve for months or even years.

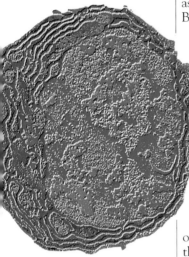

cell
his false-color electron micrograph
ows a mature B cell. About
) percent of all the lymphocytes
rculating in the blood are B cells.

Cellular immunity

form of immunity that works
y using cells

Cellular immunity is also known
s **cell-mediated immunity**. This
ranch of the immune system
ses cells that travel through the
ody to attack invaders directly.
Cellular immunity is used
gainst many kinds of viruses,
ut only a few kinds of bacteria.
is also used to destroy cells
ffected by cancer ■.

T cell

A cell used to attack invaders

T cells are lymphocytes that are
involved in cellular immunity.
Instead of making antibodies,
most T cells attack invaders
directly. There are three main
kinds of T cell, and each works
in a slightly different way. **Killer
T cells** seek out infected cells in
the body, and lock onto them.
They then release chemicals
called **lymphokines** which
destroy the infected cells,
together with the pathogens that
they contain. **Helper T cells**
assist killer T cells, and also help
B cells to produce antibodies.
Memory T cells do not
participate in the immune
response. Instead, they
memorize antigens so that
an invader can be dealt
with quickly if it appears
again in the future.

Autoimmunity

A disease in which the
body produces antibodies
against itself.

Normally, the immune system
only attacks cells or substances
that come from outside the body.
But sometimes it confuses the
body's proteins with antigens,
and starts attacking them. This is
called autoimmunity, and it can
result in an **autoimmune disease**.

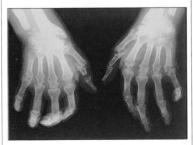

Rheumatoid arthritis
This false-color X-ray shows the hands
of someone with rheumatoid arthritis.
This autoimmune disease makes joints
stiff and inflexible.

AIDS

A disease in which the immune
system is disabled by a virus

AIDS is an abbreviation for
**Acquired Immune Deficiency
Syndrome**. It is caused by HIV,
or **Human Immunodeficiency
Virus**. This virus attacks the
immune system's helper T cells.
It reduces the number of these
cells, making it more difficult for
the body to defend itself against
attack. As a result, people with
HIV are at risk from infections
that the body can normally fight
off. AIDS first came to medical
attention in the 1980s. It has
become widespread in many
parts of the world, including
Europe, North America, and
Africa. As yet, there is no cure.

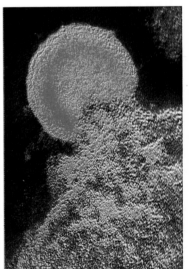

AIDS virus
This false-color electron micrograph
shows the emergence of a single HIV
particle from an infected T cell.

See also

Continued over page ➤

Immunization

The process of giving someone immunity to a disease

Immunization is also known as vaccination. It is a way of preparing the body's immune system so that it is ready to fight off a particular disease. There are two kinds of immunization. In **active immunization**, a person is given a vaccine containing pathogens ■ that have been altered so they cannot cause disease. The body reacts by making antibodies ■ against the pathogens. This protects the body from further attack. In **passive immunization**, a person is given antibodies against pathogens taken from the body of an infected person or animal. This kind of immunity does not last as long as active immunity, because the antibodies gradually break down.

Active immunization
Pathogens used in active immunization have been altered, or modified, to stop them from causing disease.

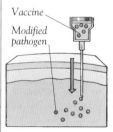

Vaccine
Modified pathogen

1 *To immunize a person against a particular disease, a vaccine containing a small amount of the pathogens that cause the disease is injected into the body.*

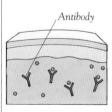

Antibody

2 *The immune system produces antibodies against the modified pathogen. It also develops memory cells.*

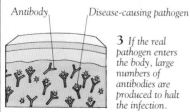

Antibody   Disease-causing pathogen

3 *If the real pathogen enters the body, large numbers of antibodies are produced to halt the infection.*

Vaccine

A medication used to produce immunity against a disease

Vaccines are used to give people active immunization. They usually contain pathogens, such as bacteria or viruses, although some contain antigens ■, such as bacterial toxins. The pathogens in a vaccine are killed, or **attenuated**, which means that they have been specially treated to make them harmless. A vaccine is usually given in an injection, called a **vaccination**, or an **inoculation**, because many disease-causing organisms are digested if they are swallowed.

Allergy

An excessive immune response to an antigen

If you are allergic, or **hypersensitive**, to something, your immune system produces an **allergic reaction**. It attacks the substance that causes the allergy, and releases substances called histamines ■ that disrupt your body's systems. Allergic reactions are triggered by substances called allergens. They can cause a wide range of symptoms ■, from skin rashes or sneezing to asthma and even loss of consciousness. Some kinds of allergic reaction occur almost instantly, but others take several hours to become noticeable.

House-dust mite
This picture shows part of a house-dust mite, a microscopic animal that lives in the house dust on mattresses and pillows. Many people are allergic to the proteins found in the feces of these creatures.

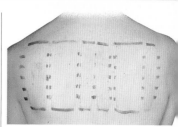

Allergen testing
Small amounts of different allergens ar applied to the skin. A slight swelling indicates that a person may be allergic to a particular allergen.

Allergen

A substance that triggers an allerg

An allergen is a substance that i harmless to most people, but harmful, or **allergenic**, to others Allergens include a huge range of different substances that we come into contact with in different ways. Some allergens produce a reaction when they come into contact with the skin These include chemicals such as perfumes, substances on plant leaves, and some metals. Many allergens, such as cereals, eggs, fish, and some drugs, including antibiotics ■, reach the body by being eaten. A third group of allergens are those in the air tha we breathe. These include polle and the feces of house-dust mites.

◄ *Continued from previous page*

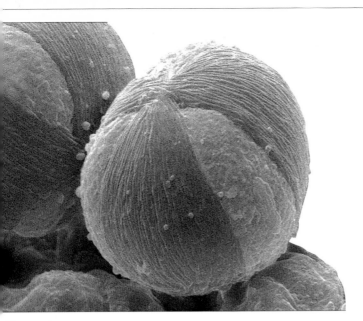

Pollen grains
Airborne pollen grains, particularly those from trees and grasses, are one of the main causes of hay fever. Hay fever is usually seasonal, and tends to occur during the summer months.

Hay fever
In allergic reaction to pollen

Hay fever is also known as **allergic rhinitis**. It is one of the commonest forms of allergy, and affects almost 10 per cent of the population. It is caused not by hay, but by tiny grains of plant pollen that drift in the air during the summer months. These pollen grains constantly settle in the eyes, and in the nasal cavity. In most people, they are washed away by tears or absorbed by mucus, and do not provoke a reaction. But in a person with hay fever, the pollen triggers an allergic reaction, which makes these parts itchy and inflamed. The inflamed cells release fluid, causing a runny nose and sneezing. The symptoms of hay fever can be relieved by **antihistamine drugs**, which counteract the effects of histamine.

Anaphylactic shock
A life-threatening allergic reaction

During this rare allergic reaction, a person's immune system releases large amounts of histamines and other chemicals. These chemicals cause widening of the blood vessels, or vasodilation ■, and this in turn leads to shock ■. Anaphylactic shock happens within minutes, and can quickly become very dangerous. It is caused by powerful allergens, such as insect stings, some drugs, and seafood.

Wasp sting
In some people, the poison in a wasp sting may cause an extreme allergic reaction, called anaphylactic shock.

Edward Jenner
English doctor
(1749–1823)

The last known case of smallpox, a deadly disease that once killed over a million people a year, occurred in 1977. Smallpox was eradicated as a result of a bold experiment by Edward Jenner, nearly 200 years earlier. Jenner heard a dairymaid say that she had suffered from a cattle disease called cowpox, so she would not catch smallpox. He collected fluid from cowpox blisters, and scraped it into a patient's skin. The patient became immune to both diseases. Jenner did not know why his treatment worked, but this pioneering experiment – the first vaccination – proved to be a breakthrough in the fight against infectious diseases.

Rejection
The destruction of foreign tissue by the immune system

During a **transplant operation**, tissues or organs are taken from one person, called a **donor**, and put into another, called a **recipient**. After a transplant, the recipient's immune system sometimes attacks the foreign cells, eventually killing them. This process is called rejection. Rejection is less common if the donor and recipient are close relatives, because they will have many antigens in common.

See also
Antibiotic 95 • Antibody 98 • Antigen 98
Histamine 94 • Pathogen 92 • Shock 89
Symptom 93 • Vasodilation 77

Metabolism

If one of your cells was enlarged to the size of a football field, you would see a swirling mass of chemicals inside it. Some of these chemicals stay much the same from one day to the next, but most are involved in complicated cascades of reactions. Together, these reactions make up your body's metabolism.

Anabolism

The part of metabolism in which substances are built up

During **anabolic reactions**, the body makes complex organic compounds out of simpler ones. These reactions normally requir a "push" in the form of energy. Examples of anabolic reactions include making proteins from amino acids ■, and making glycogen ■ from glucose.

Metabolism

All the chemical processes that take place inside the body

Even when you are fast alseep, your body carries out thousands of different chemical reactions ■ every second. Some of these reactions break down substances that you have taken in, releasing energy that keeps you alive. Others build up substances that your body needs. The two parts of metabolism – catabolism and anabolism – work side by side in cells ■. The energy released by one set of reactions is collected and used to power another set of reactions. Many metabolic reactions happen extremely quickly, because they are accelerated by special proteins ■ called enzymes ■. Their overall speed is controlled by hormones ■, which carry chemical signals from one cell to another.

See also

Amino acid 24
Body temperature 76
Carbohydrate 22 • Cell 26
Chemical reaction 21 • Digestion 116
Enzyme 24 • Excretion 77 • Fat 23
Genetic disorder 135 • Glucose 22
Glycogen 22 • Hormone 78
Organic compound 20 • Protein 24
Respiration 104 • Starch 22
Sympathetic nervous system 60
Thyroxine 80 • Urea 126

Metabolic reactions within a cell

As complex molecules are broken down during catabolism, energy is released. This energy is then used during anabolism to help build complex molecules out of simple ones.

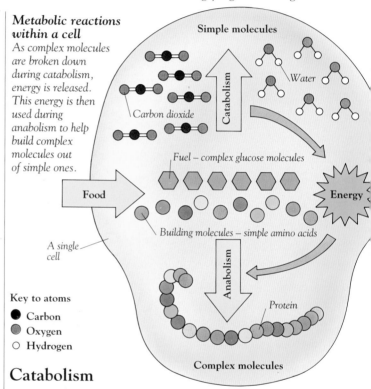

Simple molecules

Carbon dioxide

Water

Catabolism

Food

Fuel – complex glucose molecules

Energy

A single cell

Building molecules – simple amino acids

Anabolism

Protein

Complex molecules

Key to atoms

● Carbon
◐ Oxygen
○ Hydrogen

Catabolism

The part of metabolism in which substances are broken down

During **catabolic reactions**, complex organic compounds ■ are broken down to form simpler ones. These reactions usually release energy that the body can use in anabolic reactions. Some of them also provide chemical building blocks that can be used to form other substances. Examples of catabolic reactions include the breakdown of glucose ■ during respiration ■, and the breakdown of starch ■ during digestion ■.

Metabolite

A substance that takes part in the body's chemical reactions

A metabolite is any substance that is involved in metabolism. Common metabolites include glucose, which is broken down during aerobic respiration, and urea ■, which is produced by the breakdown of proteins. Some metabolites become poisonous if they are allowed to build up, so they have to be removed from the body by excretion ■.

nergy

e capacity to do work

ll living things need energy in
der to work. **Chemical energy** is
ergy stored in compounds such
fats ■ and carbohydrates ■.
hen these substances are broken
wn during catabolism,
me of this energy is
leased in the form of
at energy, which
eps you warm,
d **kinetic**
ergy, which
oves the body.
our body's energy
onstantly drains
vay to the
orld outside.
eating food,
ou make up
r this loss.

nergy expenditure
*unning converts chemical energy
to kinetic energy and heat energy.*

ilojoule

unit of energy in food

he amount of energy in food
ed to be measured in a unit
lled **calories**, or **kilocalories**
cals), but it is now given in
lojoules (**kJ**). One calorie is
ual to 4.187 kJ. Some foods, such
butter, are high in energy, while
hers, such as fruit, contain less.
ater contains no energy at all.

AVERAGE ENERGY REQUIREMENTS

Sex/age of subject	kJ/day
Infant 9–12 months	4,200
Child 8 years	8,770
Boy 15 years	12,560
Girl 15 years	9,560
Woman (inactive)	7,950
Woman (active)	9,000
Woman (breastfeeding)	11,250
Man (inactive)	10,460
Man (active)	12,560

Background picture: fruit and vegetables

Metabolic rate

The rate at which energy is released by metabolism

Metabolic rate varies constantly.
It is controlled by hormones, but
it is also affected by many
other factors including sex
and age. It is generally
higher in men than
women, and is higher
in young people than
old people. During
vigorous exercise, your
metabolic rate may
increase by up to
15 times. It also
increases by about
20 percent for every
1.8°F (1°C) rise in body
temperature ■, and it
can rise by 10–20
per cent after a meal.
Metabolic rate is
also increased by
stress, which
triggers the
sympathetic
nervous
system ■
into action.

Bar chart values (kJ/h): At rest 250, Standing 500, Walking 750, Housework 1,130, Jogging 2,640, Swimming 3,000

Metabolic rate during activity
*Metabolic rate increases as the intensity
of the exercise increases. It is shown in
kJ/h on the top of each column.*

Basal metabolic rate

The rate at which energy is released in the body at rest

Many factors affect metabolic rate,
so it is measured under standard
conditions, with a person at rest,
but not asleep. This is the basal
metabolic rate (**BMR**). It shows
how much energy the body needs
for essential processes, such as
breathing and pumping blood.
BMR is calculated by dividing the
amount of energy a person uses in
an hour by the surface area of their
body. A typical 14-year-old boy
has a BMR of about 184 $kJ/m^2/h$,
while a typical 40-year-old woman
has a BMR of about 142 $kJ/m^2/h$.

Santorio Santorio

Italian physician (1561–1636)

Santorio was
one of the first
people to use
measurements
to investigate
the way the
body works.
He found
ways to measure body
temperature and pulse rate,
but is mainly remembered for
an extraordinary and almost
lifelong investigation of
metabolism. Using a specially
designed balance, he recorded
his own weight, and the weight
of his food and body waste.
From this, he worked out that
his body lost weight through
something that could not be
seen. He called this "insensible
perspiration." We now know
that this is mainly carbon
dioxide and water vapor – two
products of respiration.

Metabolic disorder

A defect in the body's metabolic pathways

In a healthy body, metabolic
reactions are carefully regulated,
so that substances are built up or
broken down at the right time and
at the right rate. But sometimes
things go wrong. This may be
caused by a hormone being
produced in the wrong amount.
For example, in **hyperthyroidism**
too much thyroxine ■ is produced.
This speeds up the body's
metabolism, making a person
abnormally active and restless.
Metabolic disorders also occur if
an enzyme is abnormal or absent,
preventing a metabolic reaction
taking place in the normal way.
This kind of disorder is usually
caused by a defective gene ■.

Respiration

Oxygen is essential for life. Without it, your body could not obtain the energy that it needs to work. During respiration, oxygen is delivered to your body's cells, where it is used to break down fuel that comes from food.

Respiration

A chemical process in which food is broken down to release energy

The word respiration can be confusing, because it is often used to mean two different parts of the same overall process. **Cellular respiration**, or **internal respiration**, takes place inside individual cells ■. In this kind of respiration, food substances such as glucose ■ are broken down by a series of chemical reactions, most of which need oxygen. The reactions release energy which the body can then use. **External respiration** delivers oxygen to the body's cells so that internal respiration can take place. This kind of respiration involves breathing ■, which is carried out by the respiratory system ■.

Anaerobic respiration

A kind of respiration that does not involve oxygen

During anaerobic respiration, glucose is partly broken down and a small amount of energy is released. This kind of respiration takes place in a cell's cytoplasm ■, and it does not require any oxygen. If oxygen is available, anaerobic respiration is normally followed by aerobic respiration. This completes the job of breaking down glucose and releases much more energy. But if oxygen is in short supply, this second stage cannot take place.

Aerobic respiration

A kind of respiration that requires oxygen

Aerobic respiration is the body's main way of obtaining energy. It takes place in a cell's "power stations," or mitochondria ■, and it uses oxygen to break down the remains of glucose to form two waste products – carbon dioxide and water. This kind of respiration releases a large amount of energy. About 40 percent of the energy is collected and used by the body. The rest is released as heat.

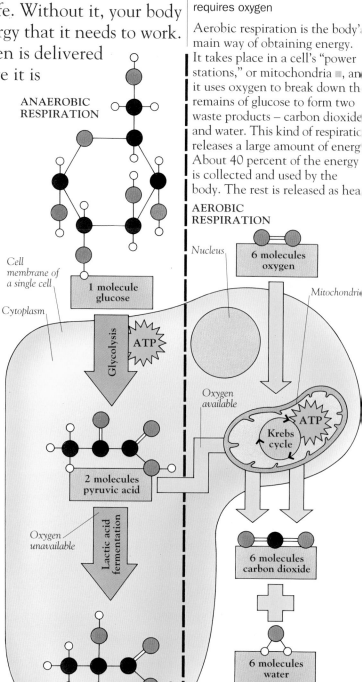

ANAEROBIC RESPIRATION

Cell membrane of a single cell

Cytoplasm

1 molecule glucose

Glycolysis

ATP

2 molecules pyruvic acid

Oxygen unavailable

Lactic acid fermentation

2 molecules lactic acid

AEROBIC RESPIRATION

Nucleus

6 molecules oxygen

Mitochondrion

Oxygen available

ATP

Krebs cycle

6 molecules carbon dioxide

6 molecules water

Key to atoms
● Carbon ● Oxygen ○ Hydrogen

Respiration
This diagram shows how glucose is broken down in a cell. Anaerobic respiration is shown on the left, and aerobic respiration is on the right.

ATP

The body's main energy carrier

ATP, or **adenosine triphosphate**, is the body's shuttle service for energy. During respiration, energy released as glucose is broken down. This energy is used to combine a compound ■ called **adenosine diphosphate (ADP)** and an ion ■ called phosphate to form ATP. ATP stores the energy and carries it to where it is needed. When an ATP molecule is broken down, it releases the energy that it has stored. The energy is used for many things, from helping cells to divide ■ to powering muscles ■.

Glycolysis

A series of chemical reactions that splits glucose molecules in two

During glycolysis, one glucose molecule with six carbon atoms is turned into two molecules of a compound called **pyruvic acid**, each of which has three carbon atoms. Glycolysis does not need oxygen, but if oxygen is available, these molecules are then broken down aerobically in the Krebs cycle. If not, the pyruvic acid is converted into a compound called **lactic acid**. This is broken down when oxygen becomes available once more.

Krebs cycle

A cycle of chemical reactions that releases energy

The Krebs cycle is at the heart of aerobic respiration. During the Krebs cycle, a molecule of pyruvic acid is gradually dismantled, so that its energy is released. It takes place in a cell's mitochondria, and needs oxygen to work. The cycle was discovered by the German biochemist Hans Krebs ■. It is also known as the **citric acid cycle**, because this acid is formed at the beginning of the cycle.

Lactic acid fermentation

The form of anaerobic respiration that occurs in the human body

During vigorous exercise, your muscles soon begin to run short of oxygen. When this happens, they can no longer carry out aerobic respiration. Instead, they break down glucose anaerobically, turning it first into pyruvic acid, and then into lactic acid. Most of the lactic acid flows out of the muscles into the bloodstream, and is dealt with by the liver ■. But some remains in the muscles, and it finally builds up and stops them from working normally. Lactic acid is the substance that makes muscles ache after they have been working hard.

Antoine Lavoisier

French chemist (1743–94)

Lavoisier was one of the most brilliant scientists of his age. In the 1770s, he discovered that when something burns, it combines with oxygen and produces carbon dioxide as well as releasing heat. Lavoisier wondered if a similar process occurred inside animals during respiration. In the 1780s, he measured the amount of carbon dioxide produced by a guinea pig, and the amount of heat energy it released. His results showed that respiration, like burning, is a form of oxidation ■, in which food is used as fuel.

Oxygen debt

The amount of oxygen needed to break down lactic acid

After hard exercise, muscles contain a lot of lactic acid. This needs to be broken down by aerobic respiration before the muscles can start working normally again. The amount of oxygen needed for this process is called an oxygen debt. Panting helps to supply the muscles with the oxygen that they need.

ANAEROBIC RESPIRATION

Glucose → Glycolysis → Pyruvic acid → Lactic acid fermentation → Lactic acid

2 ATP

Net total: 2 ATP

AEROBIC RESPIRATION

Glucose → Glycolysis → Pyruvic acid → Oxygen

8 ATP

Net total: 38 ATP

Krebs cycle — 30 ATP

Energy yield of respiration
In anaerobic respiration, for each molecule of glucose broken down, 2 molecules of ATP are formed. During aerobic respiration, the same amount of glucose in the presence of oxygen releases 38 molecules of ATP.

See also

Nutrition & diet

The body needs a steady supply of raw materials in order to survive. To meet this need, you will probably eat about 20 tons of food during your life, and drink between 4,400 and 8,800 gallons (20,000 and 40,000 liters) of water.

Nutrition

The process of obtaining the raw materials needed to stay alive

To work normally, the human body needs a regular supply of nutrients. If you go without food or water for more than a few hours, a part of your brain called the hypothalamus ■ produces the sensations of **hunger** and **thirst**. These warn you that your body needs nutrients, and prompt you to eat or drink. A healthy adult may survive without food for several weeks, but will die after a few days without water.

Nutrient

Any substance needed to maintain life

Nutrients supply your body with energy, and with the raw materials that it needs to grow and to repair itself. Foods usually contain a variety of nutrients. Most of these have to be digested ■ before they can be used. Food also contains water and dietary fiber, which are usually classified as nutrients.

See also

Amino acid 24 • Carbohydrate 22
Cell membrane 28 • Cellulose 22
Deficiency disease 108 • Digestion 116
Fat 23 • Hypothalamus 65
Inorganic compound 21 • Intestine 124
Mineral 108 • Organic compound 21
Protein 24 • Saturated fat 23
Unsaturated fat 23 • Vitamin 108

Bread

Rice   Pasta   Potatoes

Sources of carbohydrate
The two main kinds of carbohydrate are starch and sugar. Starches are complex carbohydrates found in foods such as pasta, potatoes, bread, and rice. Sugars are simple carbohydrates found in fruit, and in foods such as cakes and sweets.

Corn oil   Olive oil   Peanut oil

Butter

Sources of fat
There are two main kinds of fat – saturated and unsaturated. Saturated fat is found mainly in dairy and animal fats, such as butter. Unsaturated fat is found in most plant oils.

Macronutrient

A nutrient that is needed in large amounts

Macronutrients make up most of our food. There are three types of macronutrient – carbohydrates ■, fats ■, and proteins ■. The body's main source of energy is carbohydrates. Fats also provide energy, and are used to make cell membranes ■ and some hormones. Proteins provide amino acids ■, which the body uses to make proteins of its own.

Micronutrient

A nutrient that is needed only in small amounts

There are two kinds of micronutrient – vitamins ■ and minerals ■. They are needed only in small quantities, but they play an essential part in the body's chemistry. Vitamins are organic compounds ■ made by plants or animals. Minerals are inorganic compounds ■ that come from a variety of sources. Some are present in food, while others come from tap water or table salt.

Diet

The type and amount of food that a person eats

The word diet is often used in association with losing weight. In fact, a diet is any kind of food intake over a period of time. Diets vary greatly over the world. They depend partly on the foods that are available, and partly on what people like to eat. For good health, a diet has to be **balanced**, which means that it contains a range of the nutrients that the body needs. A diet that has too much or too little of a nutrient can cause diseases. For example, a heart attack may be caused by a diet that contains too much fat.

Dietary fiber

Plant food that cannot be digested

All plants contain cellulose ■, a carbohydrate that we cannot digest. When we eat plant food, this and other substances pass straight through the digestive system, forming dietary fiber, or roughage. Fiber adds bulk to undigested food. This gives the intestinal ■ muscles something to work against, and improves their efficiency. There are two types of dietary fiber. **Insoluble fiber** passes through the intestines unchanged. **Soluble fiber** is partly broken down by bacteria in the digestive system.

Omnivore

A person who eats both plant and animal food

Most people are omnivores, which means that they eat food from both plant and animal sources. Meat is a valuable source of protein, and also of vitamins and minerals. However, some kinds of meat contain large amounts of saturated fat ■, which can cause health problems if eaten in large quantities. By contrast, fish – a good source of animal protein – is high in unsaturated fat ■.

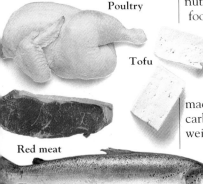

Poultry

Tofu

Red meat

Fish

Sources of protein
Meat, poultry, and fish are rich in animal proteins. Tofu contains plant proteins.

Pasta with vegetables

Sorbet

Bean salad

A vegetarian meal
A vegetarian diet is often very healthy, as it is low in fat and rich in fiber from vegetables, beans, lentils, and grains.

Vegetarian

A person who does not eat meat

There are millions of vegetarians in the world, showing that it is quite possible to live healthily without meat. Some vegetarians eat eggs and dairy products, which help to maintain their intake of proteins, vitamins, and minerals. Other vegetarians, called **vegans**, avoid animal products completely. Vegans have to eat a mixture of plant food to ensure they get all the nutrients that they need.

Malnutrition

A state of poor nutrition

Malnutrition can be caused by a diet that is lacking in a particular nutrient, or by an overall lack of food. If a person's diet is short of a micronutrient, such as iodine or vitamin C, they will eventually suffer from a deficiency disease ■. If a person is short of a macronutrient, such as protein or carbohydrate, they will lose weight, and many of their body systems will stop working normally. Malnutrition affects a large part of the world's population. Many people die from malnutrition every year, particularly in the developing (nonindustrialized) world.

Obesity

A condition in which body weight is increased by excess fat deposits

Fat deposits make up 16–25 percent of a healthy person's body weight. This fat acts as an energy store, and helps to keep the body warm. Normally, this store of fat remains fairly constant, because the body takes in as much food energy as it uses up. However, if someone takes in more energy than they use, they risk developing extra fat deposits, and becoming **obese**. Obesity may be the result of overeating, or of getting too little exercise. Occasionally, it is produced by disorders involving hormones.

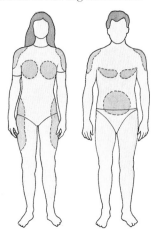

Fat deposits
Fat tends to accumulate in different areas in men and women, shown in blue.

Anorexia

A loss of appetite

Many people lose their appetite temporarily if they are anxious, tired, or ill. **Anorexia nervosa** is a much more serious condition; the sufferer refuses to eat and becomes very thin. It is believed to be a psychological disorder affecting mainly teenage or young adult women, but it can also be caused by brain or stomach tumors, strokes, or gastritis.

Vitamins & minerals

Vitamins and minerals are essential nutrients that play an important part in the body's chemistry. We need them only in small amounts, but a shortage of just one can cause ill health.

See also

Enzyme 24 • Inorganic compound 21
Metabolism 102 • Organic compound 21

Vitamin

An organic compound that the body needs in small amounts

Vitamins are organic compounds ■. Most cannot be made by the body. They are made by other living things, and we obtain them by eating food. There are more than 12 vitamins, soluble either in fat or in water. All are vital for the body's metabolism ■.

Mineral

An inorganic compound needed by the body

Minerals are inorganic compounds ■ that we obtain from food, salt, or tap water. Some are needed in moderate quantities; others, called **trace minerals**, are needed in tiny amounts. Minerals are essential for the function of many enzymes ■.

Deficiency disease

A disease produced by a shortage of a nutrient

A deficiency disease develops if the body lacks a vitamin, mineral, or other nutrient, preventing the body's chemistry from working normally. One of the best studied vitamin deficiency diseases is scurvy, which is caused by a lack of vitamin C. Sailors often used to suffer from scurvy because they lacked fresh food. By accident, fruit was discovered to be a cure.

KEY MINERALS

Mineral	Source	Function	Deficiency symptoms
Calcium (Ca)	Dairy products, green vegetables, seafood, nuts, tap water	Helps build bones & teeth; involved in nerve action	Stunted growth; rickets; osteoporosis; convulsions
Chlorine (Cl)	Table salt, seafood, milk, meat, eggs	Maintains balance of ions in body; forms acid in stomach	Muscle cramps; mental apathy; reduced appetite
Copper (Cu) *	Liver, meat, fish, cereals, mushrooms, tap water	Involved in bone formation & hemoglobin production	Anemia
Fluorine (Fl) *	Seafood, sea salt, tap water	Strengthens bones & teeth	Tooth decay (dental caries)
Iodine (I) *	Fish, shellfish, sea salt	Essential thyroid hormone	Reduced metabolic rate; swollen thyroid gland (goiter)
Iron (Fe) *	Red meat, liver, green vegetables, grains, nuts	Essential part of hemoglobin	Anemia
Magnesium (Mg)	Meat, green vegetables, whole-grain cereals	Helps build bones; enables nerves to function	Failure to grow; behavioral disturbances; weakness
Manganese (Mn) *	Vegetables, nuts, grains	Activates many enzymes	Poor growth
Phosphorus (P)	Meat, milk, dairy products, fish, cereals	Helps build bones; essential part of DNA & ATP	Weak or malformed bones
Potassium (K)	Meat, milk, cereals, fruit, vegetables	Maintains balance of ions in body; used by nerves	Muscle weakness
Sodium (Na)	Most foods except fruit	Maintains balance of ions in body; used by nerves	Muscle cramps; mental apathy; reduced appetite
Sulphur (S)	Meat, eggs, milk, nuts	Essential part of some proteins	Impaired protein synthesis
Zinc (Zn) *	Meat, eggs, fish, cereals	Essential part of some enzymes; promotes healing	Growth failure; lack of sexual maturation; loss of appetite

Background picture: crystals of table salt

Note A trace mineral is indicated by *

FAT-SOLUBLE VITAMINS

itamin	Source	Function	Deficiency symptoms
etinol)	Green & yellow vegetables, fish oil, egg yolk, liver, milk	Important in growth & formation of teeth & bones; used in vision; helps prevent infection	Night blindness; dry scaly skin; lowered resistance to infection
alciferol)	Fish oil, egg yolk; also produced in skin during exposure to sunlight	Regulates use of phosphate & calcium in bone formation; aids absorption of calcium from food	**Rickets**, a disease in which bones & other hard parts of body fail to grow properly
lpha tocopherol)	Green vegetables, plant oils, whole-grain cereals, liver	Essential for formation of red blood cells; enables some enzymes to function; prevents breakdown of fatty acids in cells	Breakdown of red blood cells
	Green vegetables; also made by bacteria in intestines	Involved in production of chemicals that enable blood to clot	Failure of blood clotting system, sometimes leading to bleeding from surfaces of body

ackground picture: spinach leaf

WATER-SOLUBLE VITAMINS

itamin	Source	Function	Deficiency symptoms
1 hiamine)	Whole grains, liver, peas, beans, yeast, nuts	Essential for functioning of enzymes that promote breakdown of carbohydrates; helps nerves & muscles function normally	**Beriberi**, a disease causing weakness & inflammation of nerves
2 iboflavin)	Milk, leafy vegetables, eggs, cheese; also made by bacteria in intestines	Helps form enzymes that control buildup & breakdown of carbohydrates & proteins	Cracked skin; defective vision
iacin	Lean meat, wheat germ, cereals, fish, yeast	Helps form enzymes that control respiration	**Pellagra**, a disease causing skin disorders & diarrhea
6 yroxidine)	Whole grain cereals, liver, egg yolk	Helps form enzymes that break down fatty acids & amino acids	Anemia; convulsions
12 yanocobalamin)	Liver, kidney, fish, eggs, milk, meat, oysters	Helps form enzymes involved in making proteins; promotes formation of red blood cells & use of carbohydrates	Anemia; impaired function of nervous system
antothenic acid	Meat, whole-grain cereals, nuts, vegetables, eggs, yeast	Helps form enzymes involved in breakdown of carbohydrates & fats; involved in nerve function & production of sex hormones	Disorders of nervous & digestive systems
olic acid folate)	Green leafy vegetables, liver, wheat germ, fruit, yeast	Helps form enzymes involved in making nucleic acids; plays a part in manufacture of red blood cells	Anemia; sores in mouth
iotin	Liver, eggs, milk, whole grains, yeast; also made by bacteria in intestines	Helps form enzymes involved in making & breaking down fats & carbohydrates	Fatigue; depression; nausea; skin disorders
ascorbic acid)	Citrus fruits, tomatoes, potatoes, leafy vegetables	Promotes formation of collagen; promotes growth of bones, teeth & blood vessels; essential for normal functioning of many enzymes; aids healing of wounds	Swollen gums & nosebleeds; severe deficiency produces the disease **scurvy**, which causes internal bleeding & swollen joints

ackground picture: slices of lemon – a citrus fruit

Respiratory system

The body needs to burn chemical fuel to release energy. This process requires oxygen. The respiratory system carries oxygen from the air into the bloodstream, and expels carbon dioxide from the body.

Respiratory system

A body system that collects oxygen from the air and expels carbon dioxide

The respiratory system works with the circulatory system ■ to deliver oxygen to the body's cells, so that aerobic respiration ■ can take place. It also enables the body to get rid of carbon dioxide, which is a waste product. The system has two parts. The respiratory tract carries air into and out of the body, while the lungs allow gas exchange ■ between the air and the blood ■.

Respiratory system
The main organs of the respiratory system, and their location within the body, are shown here.

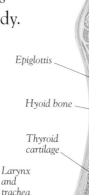

Nasal cavity

Mouth

Right lung

Larynx and trachea

Left lung

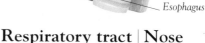

Epiglottis

Hyoid bone

Thyroid cartilage

Larynx and trachea

Adam's apple

Trachea

Esophagus

Fro s...

N. ca...

N...

Max (up j...

Ton...

Mand (lower j...

Nose, mout and thro
This model sho the upper part of respiratory syst in cross sectio

See also

Respiratory tract

The pathway that carries air to and from the lungs

The respiratory tract allows air that is inhaled ■ to reach the lungs. Air usually flows in and out of the body through the nose, but during strenuous exercise, when more air is needed, it flows through the mouth. The first part of the tract is the nasal cavity. From here, the air travels through connected passages until it reaches microscopic air sacs called alveoli ■ inside the lungs. A membrane ■ lining most of the respiratory tract produces sticky mucus, which traps foreign objects such as bacteria ■ and dust. The mucus is moved away from the lungs by tiny hairs called cilia ■. It is either swallowed, or expelled from the body when you cough, sneeze, or blow your nose.

Nose

An organ used for breathing and smelling

The nose has two entrances, called **nostrils**. These are separated by a partition called the **nasal septum**, which is mad of cartilage ■. Each nostril is lined with hairs that help to prevent dust and dirt from getting into the respiratory trac

Nasal cavity

A hollow space behind the nose

When you breathe in through your nose, air flows through an air-filled space called the nasal cavity, where it is warmed up. Like the outer part of the nose, the nasal cavity is divided in tw by a partition, or septum, which in this case is made of bone. Th cavity is lined with membranes that produce mucus, and is connected to several smaller spaces, called sinuses.

harynx

passage that carries air and food

he pharynx is more commonly nown as the **throat**. Its upper rt carries air, and is connected the nasal cavity by two penings that are each about the e of a fingernail. It also has o smaller openings. Each of ese is the lower end of a ustachian tube ■, which lets air ach the middle ear ■. The iddle part of the pharynx nnects with the back of the outh, and it carries food as well air. In the lower part of the harynx, the passage divides. od continues downwards into e esophagus ■, but air flows rward and through the larynx.

arynx

chamber made of cartilage

he larynx is also called the ice box. It is a complex rangement of pieces of rtilage. Together, these form a ort funnel that links the se of the throat with e trachea. The rynx can be osed off by the pper piece of rtilage, the iglottis. The wer piece of rtilage, the icoid cartilage, taches the rynx to the trachea. between are the yroid cartilage and e arytenoid cartilages, hich hold the vocal rds in position. From e front, the larynx is ughly triangular, and has forward-pointing bump lled the **Adam's apple**.

he larynx
veral pieces of cartilage make up e larynx. This side view shows w they are arranged.

Epiglottis

A leaf-shaped flap of cartilage at the back of the larynx

The epiglottis folds forwards when you swallow to prevent food or liquids from ending up in the lungs. It seals off the larynx until the food or liquid has entered the esophagus, and is safely out of the way.

Choking

An inability to breathe caused by a blockage in the respiratory tract

The epiglottis is not always completely effective, and things sometimes "go down the wrong way" when you swallow. When this happens, a strong reflex action occurs that causes a bout of coughing ■. This usually clears up the problem. Sometimes, a small object, such as a piece of food, can get stuck in the airways. Unless the obstruction is cleared, the result can be fatal.

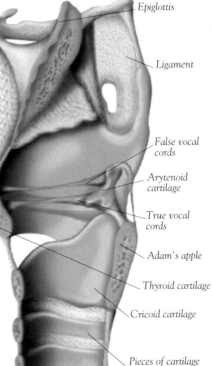

Epiglottis

Ligament

False vocal cords

Arytenoid cartilage

True vocal cords

Adam's apple

Thyroid cartilage

Cricoid cartilage

Pieces of cartilage reinforce trachea

Vocal cord

A taut membrane inside the larynx

The two pairs of vocal cords are attached to the inside of the Adam's apple, and run backward across the larynx. The upper pair, the **false vocal cords**, shut off the larynx when you swallow. The lower pair, the **true vocal cords**, can be stretched by muscles to produce sound. The cords vibrate as the air flows past them. If they are loose, the air vibrates slowly, and produces a sound with a deep pitch. But if the cords are tight, the air vibrates much more rapidly, and produces a high note. Men usually have larger vocal cords than women, which also affects the pitch of the sound produced.

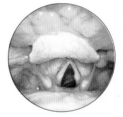

Vocal cords open
The vocal cords do not produce any sound when they are open.

Vocal cords closed
When air passes between the closed cords, they vibrate and produce sounds.

Trachea

An air pipe that leads to the lungs

The trachea is also called the **windpipe**. It begins just below the larynx, and connects with the bronchi ■ that enter the lungs. The trachea is reinforced by up to 20 pieces of cartilage. Each piece is C-shaped, with the open part of the C facing toward the back of the body. Together, these pieces of cartilage help to keep the trachea open.

Lungs

The inner surface area of your lungs is about 35 times larger than the surface area of your skin, yet is packed into a space smaller than a shopping bag. This huge area allows the body to absorb oxygen and expel carbon dioxide efficiently.

Nose

Phar

Trac

Bronc

Mouth

Right lung

Diaphra

Ril

Inside the chest
When air is breathed in, it enters the nose, travels down the trachea, throu, the bronchus, and into the lungs. The are part of the respiratory system.

Lung

A respiratory organ

The two lungs are situated on either side of the heart ■, in the chest cavity, or thoracic cavity ■. They are protected by the rib cage ■. The lungs rest on the diaphragm ■, and extend upwards to a point just above the collarbones, or clavicles ■. Each lung is a cone-shaped mass of spongy tissue with a rich blood supply. Microscopic air sacs called alveoli are grouped together in each lung in pockets called **lobules**. The lobules are grouped into **segments**, and these in turn are organized into **lobes**. The right lung has three lobes, but the left lung has just two to allow space for the heart.

The lungs
This model shows the structures of the thoracic cavity, and the position of the lungs within the cavity.

Upper lobe of right lung

Middle lobe of right lung

Lower lobe of right lung

Pleural membrane

A membrane that surrounds a lung

Each lung is enclosed by a pair of tough pleural membranes ■. These protect the lungs, and also help them to change shape. The outer membrane lines the inside of the chest; the inner membrane covers the lung itself. The narrow space between the membranes is filled with a thin film of **pleural fluid**, which lubricates the membranes, and allows them to slide past each other during breathing.

Epiglottis

Hyoid bone

Thyroid cartilage — *Larynx*

Cricoid cartilage

Trachea

Superior vena cava

Upper lobe of left lung

Aorta

Left pulmonary artery

Primary bronchus

Lower lobe of left lung

Secondary bronchus

Tertiary bronchus

Diaphragm

Bronchial tree

A branching collection of airways in the lungs

When air flows in and out of the lungs, it moves through a collection of passageways that looks like an upside-down tree. The "trunk" of the tree is the trachea ■. Its "branches" are tub called bronchi, and its "twigs" are tubes called bronchioles.

Bronchus

An airway leading from the trachea into a lung

A bronchus (plural **bronchi**) is tube that carries air into and ou of the lungs. Each lung has a single **primary bronchus**, which is reinforced by C-shaped pieces of cartilage ■. The primary bronchus divides into two or three **secondary bronchi**, each which supply air to a single lobe. The secondary bronchi divide into **tertiary bronchi**, which each supply a single segment.

Bronchiole

An airway leading from a bronchus to individual alveoli

Bronchioles are the smallest part of a lung's air distribution system. They carry air into alveoli, where gas exchange ■ takes place. Unlike bronchi, bronchioles are not reinforced by cartilage. Instead, they have a layer of smooth muscle ■. When this muscle contracts, it alters the size of the bronchiole.

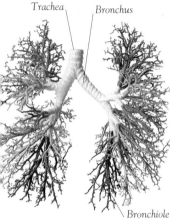
Trachea   *Bronchus*

Bronchiole

branching tree
This model shows the branching network of bronchi and bronchioles that makes up the bronchial tree.

Alveolus

A sac in which gas exchange takes place

An alveolus (plural **alveoli**) is a thin-walled sac with a moist inner surface that allows gases to be exchanged between air and blood. A cluster of alveoli looks like a tiny bunch of grapes surrounded by a network of capillaries ■. Although alveoli are small, they are extremely numerous. An adult's lungs contain over 300 million alveoli, and their total inner surface area is about 750 ft^2 (70 m^2). This large area is necessary to supply the body with the oxygen it needs, and to get rid of carbon dioxide at a fast enough rate.

Pulmonary surfactant

A substance that prevents alveoli from collapsing

The film of fluid inside an alveolus is like an open-ended bubble. Its molecules attract each other, creating a force called **surface tension**. Without surfactant, this force would make the bubble shrink, and could make the alveolus collapse. Surfactant is a mixture of phospholipids ■ and proteins. It helps to reduce surface tension, and ensures that alveoli stay inflated at all times. Premature babies sometimes have to be given extra surfactant to help them breathe.

Dust cell

A cell that engulfs dust and other particles that reach the lungs

During breathing, dust, pollen, and other particles are swept into the respiratory tract ■. Most of these are trapped by mucus or hairs, but the smallest find their way into the lungs. Here, they are dealt with by wandering phagocytes ■ called dust cells, or **alveolar macrophages**. Dust cells engulf these particles by phagocytosis, and prevent them from clogging up the lungs and interfering with gas exchange.

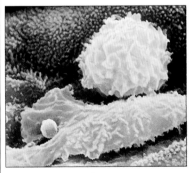

Dust cell
This false-color electron micrograph shows two dust cells inside a lung. The lower dust cell has elongated itself to engulf the small particle, bottom left.

Adolf Fick

German physiologist (1829–1901)

Adolf Fick was one of the first people to apply physics to the study of the human body. In 1856, he published an important book called *Medical Physics*, which investigated many processes that take place inside the body. These included the movement of blood inside blood vessels and the movement of heat. Today, Fick is remembered mainly for his law of diffusion, which shows how rapidly molecules spread – for example, when oxygen enters the bloodstream through the lungs.

Respiratory membrane

A membrane that separates air and blood

Inside the lungs, air and blood are separated by a membrane that is just two cells thick. This membrane is made up by the wall of the alveolus, and the wall of the adjoining capillary. The total distance across the membrane is often less than 0.00004 inch (0.001 mm), and gases can easily travel across it by diffusion ■. The side facing the air is covered with a film of fluid. This dissolves the oxygen before it travels across the membrane and into the blood.

See also

Breathing & gas exchange

You breathe continuously, drawing fresh air into your lungs and forcing stale air out of your body. The air that you inhale contains oxygen, which passes from your lungs into your blood and around your body. The air that you exhale travels in the opposite direction and contains carbon dioxide.

Breathing

The movement of air into and out of a lung

In order to survive, humans need to take in oxygen and expel carbon dioxide waste in large amounts. This is brought about by pumping air into and out of the lungs in a process called breathing. The lungs do not have any muscles of their own. Air moves into or out of them as a result of changes in the shape of the thoracic cavity ■. These changes are produced mainly by the diaphragm and the intercostal muscles ■. In deep breathing, other muscles are also involved. Neck muscles increase the volume of the thoracic cavity, while abdominal muscles decrease it.

Diaphragm

A dome-shaped sheet of muscle that draws air into the lungs

The diaphragm is a muscular dome that separates the chest, or thorax ■, from the abdomen. When you breathe in, it contracts and becomes flatter. This enlarges the thoracic cavity, and draws air into the lungs. In normal breathing, the diaphragm works with the intercostal muscles, which raise the ribs.

Inhalation

The movement of air into a lung

Taking a breath is known as inhalation, or **inspiration**. During inhalation, the diaphragm contracts, and the intercostal muscles raise the ribs. Together, these two movements increase the volume of the thoracic cavity. This makes the lungs enlarge, reducing the pressure of the air inside them. The air outside the body is now at a greater pressure than the air inside the lungs. It flows through the respiratory tract ■ and into the lungs, until the air pressure inside and outside the lungs is equal.

Air is drawn into the lungs

Intercostal muscles contract and pull ribs upward and outward

Volume of lungs increases

Diaphragm contracts and moves downward

Inhaling air
The diaphragm and the intercostal muscles contract, increasing the chest cavity and letting air flow into the lungs.

Exhalation

The movement of air out of a lung

During exhalation, or **expiration**, the diaphragm and intercostal muscles relax, reducing the volume of the thoracic cavity. This compresses the lungs, squeezing the air inside them so that its pressure rises above that of the air outside. As a result, air flows from the lungs and out of the body. Exhalation is normally a passive process, but during deep breathing, such as **panting**, extra muscles contract to make the thoracic cavity even smaller. This squeezes out extra air, so that more fresh air can be taken in with the next breath

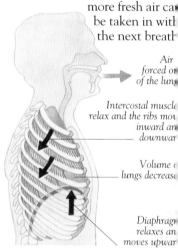

Air forced out of the lungs

Intercostal muscles relax and the ribs move inward and downward

Volume of lungs decreases

Diaphragm relaxes and moves upward

Exhaling air
During exhalation, the intercostal muscles and the diaphragm relax, and the lungs return to their normal size. This forces air out of the body.

Breathing rate

The rate at which breathing takes place

At rest, most people take about 12–15 breaths per minute, but during hard exercise, this rate can more than double. The breathing rate is controlled by the **respiratory center**, which lies in the brain stem ■. Although you can deliberately alter your breathing rate, your brain stem keeps overall control.

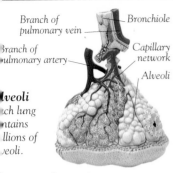

Branch of pulmonary vein — Bronchiole

Branch of pulmonary artery — Capillary network

Alveoli

Alveoli
Each lung contains millions of alveoli.

Gas exchange

The movement of gases between the air and the blood

Inside the lungs, gas exchange takes place between the air and the blood by a process called diffusion ■. Oxygen makes up about 21 percent of the air you breathe in, or **inhale**, but only about 16 percent of the air you breathe out, or **exhale**. The other 5 percent passes from the alveoli ■ through the thin walls of capillaries and into the blood. From here, it is carried around the body. At the same time, carbon dioxide passes in the opposite direction. The air that you exhale contains about 100 times more carbon dioxide than the air that you inhale.

Total lung capacity

The total volume of air in fully inflated lungs

If you breathe in as hard as you can, your lungs will contain a large volume of air. The exact figure depends on age, sex, and build, but most adults have a lung capacity of about 366 cubic inches (6 liters). The amount of air that you breathe in and out during rest is called the **tidal volume,** which is much smaller than the total lung capacity. In adults, it is about 30.5 cubic inches (0.5 liters).

Vital capacity

The volume of air that moves during deep breathing

During vigorous exercise, you breathe much more deeply than usual. Your lungs expand more when you breathe in, and they also contract more when you breathe out. The vital capacity is a measure of the total amount of air that moves during these conditions. In most people, it is about 10 times as much as the tidal volume.

See also

Alimentary canal 116 • Alveolus 113
Brain stem 64 • Diffusion 28
Intercostal muscle 53 • Reflex 63
Respiratory tract 110 • Thoracic cavity 14
Thorax 14 • Vocal cord 111

Residual volume

The minimum volume of air that always remains in the lungs

The lungs contain air spaces that cannot be emptied. This means that you can never expel all the air inside them. In adults, this residual air has a volume of about 61 cubic inches (1 liter).

Coughing

A breathing movement that clears the respiratory tract

Coughing is either a deliberate or a reflex ■ action that clears an irritation in the upper part of the respiratory tract. It begins with an unusually deep inhalation. The vocal cords ■ then close, holding the air in the lungs. At the same time, muscles make the chest contract, putting the air in the lungs under high pressure. Suddenly, the vocal cords open, and a blast of air surges out of the mouth at more than 100 mph (160 km/h). A **sneeze** is similar to a cough, but air passes through the nose rather than the mouth.

Hiccup

A sudden inhalation caused by contraction of the diaphragm

A hiccup, or **hiccough**, is the result of a sudden, involuntary contraction of the diaphragm. Air rushes into the lungs, and is shut off by the sudden closure of the vocal cords, which makes a sharp sound. Hiccups are usually triggered by nerves in the alimentary canal ■, which is why they often occur just after eating.

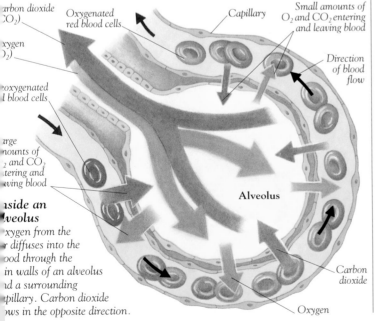

Carbon dioxide (CO₂)

Oxygen (O₂)

Oxygenated red blood cells

Deoxygenated red blood cells

Large amounts of O₂ and CO₂ entering and leaving blood

Inside an alveolus
Oxygen from the air diffuses into the blood through the thin walls of an alveolus and a surrounding capillary. Carbon dioxide flows in the opposite direction.

Capillary

Small amounts of O₂ and CO₂ entering and leaving blood

Direction of blood flow

Alveolus

Carbon dioxide

Oxygen

Digestion

We all have to eat to live, but eating is just the first stage in getting nourishment from our food. Nearly everything that we eat has to be broken down before it can be used, and this task is carried out by the digestive system.

Digestion

A process that breaks food down so that it can be used by the body

Most food consists of substances that are made up of complex molecules ■. Before your body can absorb these substances, it breaks them down into smaller units. This is carried out by proteins called enzymes ■. Enzymes work by speeding up a particular chemical reaction that makes a food substance break down. Once this has happened, the nutrients ■ from the digested food can pass into the body.

See also

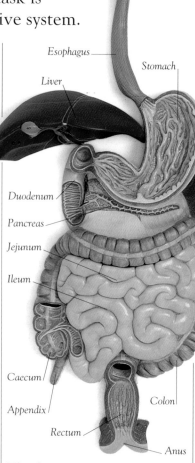

Mouth
Esophagus
Liver
Stomach
Duodenum
Pancreas
Jejunum
Ileum
Caecum
Appendix
Colon
Rectum
Anus

The digestive system
The model above shows the arrangement of the body's digestive organs. If the digestive system were unraveled and laid out in a long line it would look like it does below. This shows how long the small intestine is.

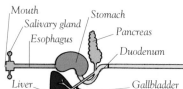

Mouth
Salivary gland
Esophagus
Stomach
Pancreas
Duodenum
Liver
Gallbladder
Jejunum

Digestive system

A system of organs that work together to digest food

The main part of the digestive system is a tube called the **gastrointestinal tract**, or **alimentary canal**. This tube runs from the mouth ■ to the anus ■, and is up to 30 ft (9 m) long. The structure of the tube changes to suit it to different purposes. For example, the esophagus ■ has very muscular walls that are used for swallowing, while the small intestine ■ is lined with villi ■ that absorb digested food into the body. The digestive system also contains other organs that assist with the break up of food, including the tongue, teeth, salivary glands ■, gallbladder ■, liver ■, and pancreas ■.

Peritoneum

A membrane that lines the abdominal cavity

The peritoneum anchors the the digestive system to the wall of the abdominal cavity ■. It contains nerves, blood vessels, and lymph vessels ■. Parts of the membrane, called **mesenteries**, hang from the abdomen and hold digestive organs in place.

Sphincter

A ring of muscle that shuts off an opening

A sphincter holds food in place temporarily, so that digestion or absorption can occur. Sphincters occur in the stomach ■, cecum ■, and anus, where they close off parts of the alimentary canal. They are also found in the circulatory system ■ and the urinary system ■.

Key
− **Sphincter**
□ **Small intestine** ■ **Large intestine**

Ingestion

The process of eating or drinking

When you ingest something, you swallow it, and start it on a journey through your alimentary canal. This journey takes up to two days. By the time it is finished, all the useful substances in the food have been digested and absorbed.

Peristalsis

Movement through a hollow organ produced by waves of muscle contraction

Most of the alimentary canal is surrounded by two sheets of smooth muscle ■. During peristalsis, these muscles work together to push food along. The inner sheet of **circular muscle** contracts behind the food, making the canal narrower. At the same time, the outer sheet of **longitudinal muscle** contracts in front of the food, making the canal wider. As a result, the food moves forward. Peristalsis is also used in other parts of the body, including the ureters ■, which carry urine to the bladder.

Peristalsis
Waves of contraction and relaxation propel food down the esophagus.

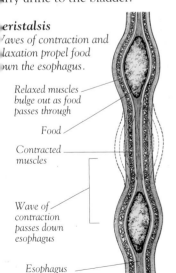

Relaxed muscles bulge out as food passes through
Food
Contracted muscles
Wave of contraction passes down esophagus
Esophagus

Carbohydrate digestion

The digestion of carbohydrates in food

Carbohydrate ■ digestion occurs mainly in the small intestine and involves the enzymes salivary amylase ■, pancreatic amylase ■, maltase ■, sucrase ■, and lactase ■. Complex chains of polysaccharides ■ are split into disaccharides ■, which are split further into monosaccharides ■. Monosaccharides have small enough molecules to be absorbed intact. Some carbohydrates cannot be digested. These form dietary fiber ■.

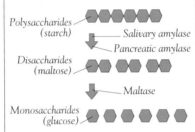

Polysaccharides (starch) — *Salivary amylase* — *Pancreatic amylase*
Disaccharides (maltose) — *Maltase*
Monosaccharides (glucose)

Digestion of carbohydrate
Complex carbohydrates are split into simpler ones so they can be absorbed.

Protein digestion

The digestion of proteins in food

Protein digestion occurs in the stomach and small intestine. It involves hydrochloric acid and the enzymes pepsin ■, trypsin ■, and peptidase ■. Protein molecules are split into small peptide chains and then into single amino acids ■ that can be absorbed.

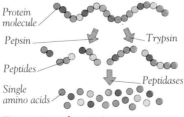

Protein molecule
Pepsin — *Trypsin*
Peptides — *Peptidases*
Single amino acids

Digestion of protein
Long protein chains are broken down into their constituent amino acids.

Fat digestion

The digestion of fats in food

Fat digestion occurs in the small intestine, and involves bile salts ■ and the enzyme lipase ■. Fats are broken down to form fatty acids ■ and monoglycerides ■. These are both in the form of tiny globules called **micelles**. Digested fats do not enter the bloodstream directly. They travel through the lining of the alimentary canal, and into lymph vessels called lacteals ■.

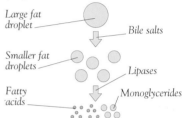

Large fat droplet — *Bile salts*
Smaller fat droplets — *Lipases*
Fatty acids — *Monoglycerides*

Digestion of fat
Fat is broken down into smaller droplets and then fatty acids and monoglycerides.

Absorption

The movement of digested food out of the alimentary canal and into the body

Strictly speaking, food in your alimentary canal is outside your body. This is because food moves through a hollow space, or **lumen**, that connects with the outside. After food has been digested, it is absorbed into the body itself. During absorption, carbohydrates and proteins travel through the walls of the alimentary canal, and into the blood, while fats enter lymph. The nutrients are then carried to where they can be used.

Assimilation

The use of nutrients by single cells

Blood carries nutrients all over the body, where they enter tissue fluid ■, and then individual cells. Here, they can be broken down during respiration ■ to release energy, or used to make new substances that the cell needs.

Ileum — *Appendix* — *Cecum* — *Colon* — *Rectum* — *Anus*

Teeth

Teeth are like a living tool kit. Their various shapes enable them to cut, crush, and grip so that they can prepare food for digestion. Teeth are harder than bones, and they develop in a very different way.

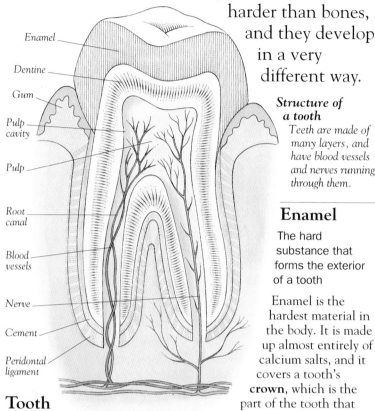

Enamel
Dentine
Gum
Pulp cavity
Pulp
Root canal
Blood vessels
Nerve
Cement
Peridontal ligament

Structure of a tooth
Teeth are made of many layers, and have blood vessels and nerves running through them.

Tooth
A hard structure in the mouth that cuts or grinds food

Teeth cut up food so that it can be swallowed ■. They also chew it, creating a larger surface area for digestive enzymes ■ to work on. Teeth form in the jaws ■. A lining of tissue called **gums** covers the jaw bones. When teeth are ready to be used, they grow out, or **erupt**, through the gums. Once a tooth has erupted, it does not get any bigger. During their lives, most people have a total of 52 teeth in two separate sets. The number, arrangement, and shape of your teeth is known as your **dentition**, and a dentist records it by writing out a **dental formula**.

Enamel
The hard substance that forms the exterior of a tooth

Enamel is the hardest material in the body. It is made up almost entirely of calcium salts, and it covers a tooth's **crown**, which is the part of the tooth that projects above the gums. Unlike all other parts of a tooth, enamel is nonliving. It stops the tooth wearing away, and protects the softer tissues inside it.

Dentine
A bonelike tissue that gives a tooth its shape

Dentine is the most abundant tissue in a tooth, and gives it its basic shape. It is not as hard as enamel, but it is still extremely strong. It forms a layer underneath a tooth's crown, and it extends downwards to form anchorages called **roots**. Dentine contains calcium salts but, unlike enamel, it also contains living cells.

Pulp
The soft tissue inside a tooth

Pulp contains blood vessels ■, lymph vessels ■, and nerves ■. It makes dentine, and keeps the inside of a tooth alive. Most pulp is in a space called the **pulp cavity**. It also extends to a tooth's roots inside spaces called **root canals**.

Cement
A hard substance that holds a tooth in place

Cement covers a tooth's roots. It contains fibers that anchor the tooth to its **socket**, which is a hollow in the jaw. The socket is lined with tissue called the **peridontal ligament**. This acts as a shock absorber when you bite.

Tooth decay
The destruction of enamel and dentine in a tooth

If teeth are not kept clean, food and bacteria ■ can build up to form a deposit called **plaque**. The bacteria release acids that eat away at the enamel and expose the inside of the tooth. This is called tooth decay, or **dental caries**. In severe cases, the interior of the tooth becomes infected and dies. Tooth decay is common in people who have a sugar-rich diet ■, and it triggers a persistent pain called **toothache**.

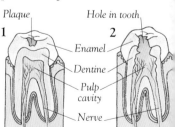

Plaque
Hole in tooth
Enamel
Dentine
Pulp cavity
Nerve

Tooth decay
(**1**) *Plaque builds up on the tooth.*
(**2**) *A hole is created in the enamel and spreads to the dentine and pulp cavity. Eventually, the nerve may be killed and the tooth may die.*

...ruption of teeth

...eth erupt in a certain order. The ...erage age of eruption of teeth in both ...ciduous (milk) and permanent (adult) ...s is shown here. All the adult teeth, ...cept the molars, push out ...e milk teeth as they ...erge. The ...ult premolars ...lace the milk ...lars. The ...ult molars ...e extra teeth.

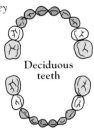

Deciduous teeth

...ey

Central incisor 6–12 months
Lateral incisor 9–16 months
Canine 16–24 months
First molar 12–16 months
Second molar 24–32 months

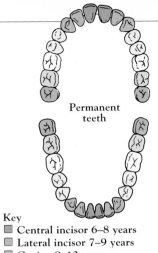

Permanent teeth

Key

■ Central incisor 6–8 years
■ Lateral incisor 7–9 years
■ Canine 9–12 years
□ Premolar 10–12 years
□ First molar 6–7 years
□ Second molar 11–13 years
■ Third molar (wisdom) 17–21 years

Deciduous teeth

The first set of teeth

...umans have two sets of teeth. ...he first set, called deciduous ...eth, or **milk teeth**, starts to erupt ... about 6 months of age, and is ...ually complete by 32 months. ...consists of 20 teeth altogether, ...ith 4 incisors, 2 canines, and ...molars in both the upper and ...wer jaws. Deciduous teeth have ...ery short roots. Before the teeth ...e pushed out and replaced by ...ermanent teeth, the roots are ...artly reabsorbed. This makes ...em easily loosened in accidents.

Replacement of deciduous teeth
The X-ray (top) shows deciduous teeth being pushed out and ...placed by permanent teeth from the ...ms below. The girl (bottom) has lost ...me of her deciduous teeth, which will ...ter be replaced by permanent teeth.

Permanent teeth

The second set of teeth

Permanent teeth, also called **adult teeth**, consist of 32 teeth. They begin to appear at about 6 years of age, and are complete by early adulthood. In each jaw, there are 4 incisors, 2 canines, 4 premolars, and 6 molars. Incisors, canines, and premolars replace deciduous teeth, but adult molars are "extra" teeth that develop as the jaw gets longer. Permanent teeth have longer roots than deciduous teeth, and are anchored very firmly.

Incisor

A tooth with a cutting edge

Incisors are at the front of the upper and lower jaws and are the first teeth to erupt in each set. They slice past each other to cut up food and have a single root.

Canine

A tooth that grips and pierces

Canine teeth, or **eye teeth**, are on either side of the incisors. They have a single root. Human canines are smaller than the equivalent teeth in dogs, from which they get their name.

Premolar

A tooth that is used for crushing and chewing

Premolars are also called **bicuspids**, because they have two raised edges, or **cusps**. They are positioned behind the canines, and have one or two roots. Adults have 8 premolars that replace the 8 molars in the first set of teeth.

Molar

A tooth that crushes and chews with maximum force

Molars are at the back of the jaw. They have four cusps and, in adults, two or three roots. Because they are near the hinge of the jaw, molars bite with tremendous force, and can crush most kinds of food. Permanent molars appear gradually, because the jaws have to grow longer to make room for them. The **first molars**, nearest the front of each jaw, appear at about 6–7 years of age, followed by the **second molars** at about 11–13 years.

Wisdom tooth

A molar in the angle of the jaw

Wisdom teeth are also known as **third molars**. They usually appear after the age of 17 years, but in some people they never erupt at all. If the teeth do appear, they erupt into a narrow space at the back of the jaw. If the space is too small, wisdom teeth often become **impacted**, which means that they remain embedded in the jaw. Impacted wisdom teeth can cause severe pain, and they sometimes have to be removed.

See also

Bacteria 92 • Blood vessel 88
Diet 106 • Enzyme 24 • Jaw 39
Lymph vessel 96 • Nerve 58
Swallowing 120

Mouth & esophagus

Your mouth is the entrance to your digestive system, as well as a passageway that carries air and sound waves. You can control its movements, so that you bite and chew with just the right force. But once food has entered your esophagus and is on its way to your stomach, the rest of digestion runs automatically.

Mouth

The entry to the alimentary canal

The space inside the mouth is called the **buccal cavity**, or **oral cavity**. Like most parts of the alimentary canal ■, the mouth contains many bacteria ■. Inside the mouth, food is crushed by the teeth, and mixed with saliva before being swallowed. The tongue ■ plays a part in eating, tasting, and talking. Taste buds ■ on the tongue's surface sense substances in food. When you talk, you use your vocal cords ■ to make basic sounds, but your tongue modifies this sound by altering the space inside your mouth.

Palate

The roof of the mouth

The roof of the mouth is hard at the front and soft at the back. These areas are called the **hard palate** and the **soft palate** respectively. The hard palate is a bony arch covered by a mucous membrane ■. The soft palate consists of muscle covered by the same kind of membrane. Attached to this is a small strip of muscle called the **uvula**, which hangs in the entrance to the throat, or pharynx ■. The palate divides the oral cavity from the nasal cavity ■.

Salivary gland

A gland that produces saliva

Saliva is a digestive fluid that helps food to slip down the throat. It contains the enzyme ■ **salivary amylase**, which digests starch ■. Saliva is produced mainly by three pairs of glands that connect to the mouth by short channels, or ducts. The **parotid glands** are in the cheeks, the **sublingual glands** are just below the tongue, and the **submandibular glands** are near the back of the mouth, below the tongue. Saliva is created all the time, but large amounts are released when you see or smell food. This sudden rush of saliva makes your mouth "water."

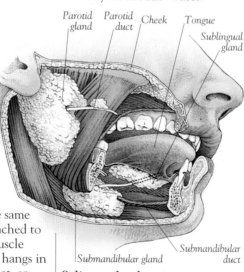

Parotid gland   Parotid duct   Cheek   Tongue   Sublingual gland

Submandibular gland   Submandibular duct

Salivary glands
The mouth has three pairs of salivary glands situated around it.

Esophagus

A muscular tube that leads from the mouth to the stomach

The esophagus, or **gullet**, carries food from the mouth to the stomach by peristalsis ■. Unlike the trachea ■ in front of it, it does not have reinforced walls and it is pressed flat when not in use.

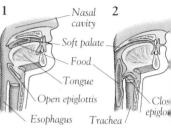

1   Nasal cavity   2

Soft palate

Food

Tongue

Open epiglottis   Closed epiglottis

Esophagus   Trachea

Swallowing
(1) Food is pushed to the back of the throat and (2) enters the esophagus. At this moment, the epiglottis closes to stop food entering the trachea, and the soft palate rises to block the nasal cavity.

Swallowing

A series of movements that carries food or liquid into the stomach

Swallowing involves voluntary and automatic movements. The tongue voluntarily moves food to the back of the mouth. The food enters the throat, and triggers a wave of peristalsis that carries it down the oesophagus. At the same time, the epiglottis ■ automatically closes the trachea, stopping food entering the lungs. During **vomiting**, this process is reversed, and the contents of the stomach are emptied. This reflex action is an important defense against harmful substances, but is also triggered by many diseases.

See also

Alimentary canal 116 • Bacteria 92
Enzyme 24 • Epiglottis 111
Mucous membrane 19 • Nasal cavity 110
Peristalsis 117 • Pharynx 111
Reflex 63 • Starch 22 • Taste bud 75
Tongue 75 • Trachea 111 • Vocal cord 111

Stomach

Your stomach is one of the most elastic organs in your body. Once it has filled with food, digestion begins. Although it contains the most powerful acid in the body, a special lining ensures that it digests food, but does not digest itself.

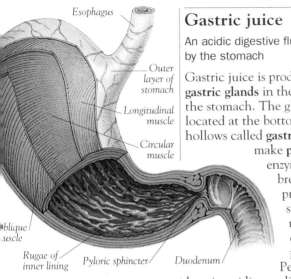

Esophagus

Outer layer of stomach

Longitudinal muscle

Circular muscle

Oblique muscle

Rugae of inner lining

Pyloric sphincter

Duodenum

Structure of the stomach
The stomach has three layers of muscle. They contract in different directions to thoroughly break up and mash food.

Stomach

A curved chamber in the alimentary canal

The stomach lies beneath the diaphragm ■, at the top of the abdominal cavity. Its upper end is connected to the esophagus ■, and its lower end to the duodenum ■. The stomach stores food, and produces a powerful digestive juice that breaks down proteins. Its inner walls are thrown into deep folds called **rugae**, which stretch when the stomach fills with food. The walls are very muscular, and they churn up food so that it can be digested. The muscle layer is divided into the longitudinal, circular, and oblique layers.

Gastric juice

An acidic digestive fluid produced by the stomach

Gastric juice is produced by **gastric glands** in the lining of the stomach. The glands ■ are located at the bottom of deep hollows called **gastric pits**. They make **pepsin**, an enzyme ■ that breaks down proteins into smaller molecules called peptides ■. Pepsin works best in acidic conditions, so the glands release **hydrochloric acid**. The acid also kills bacteria ingested with the food. The stomach does not digest itself because pepsin does not work until it mixes with the acid, and also because the glands make a mucus ■ that acts as a protective lining.

Gastric pits
This micrograph shows the lining of the stomach. The dark hollows are gastric pits, at the bottom of which are gastric glands. The glands produce gastric juice to help break down proteins.

Rennin

An enzyme that coagulates milk

Rennin is produced in the stomach of infants, but not of adults. It turns milk ■ into a loose solid, preventing it from leaving the stomach too quickly. This ensures that it remains there long enough for pepsin to act on it.

Chyme

A liquid containing partly digested food

Chyme forms after the stomach has mixed up food and begun to digest it. Once the food has been liquefied, it is ready to pass into the duodenum.

Pyloric sphincter

A ring of muscle that seals off the base of the stomach

When you eat a meal, your pyloric sphincter closes the tube that connects your stomach to your duodenum. After the food has been partly digested, the sphincter opens slightly and begins to let chyme flow through. Foods rich in fat ■ or protein may stay in the stomach for three hours, whereas foods rich in carbohydrates ■ usually stay less than an hour. The **lower esophageal sphincter**, at the end of the esophagus, controls the amount of food entering the stomach.

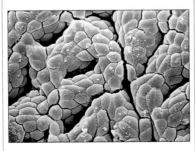

Pyloric sphincter
This sphincter controls the passage of food from the stomach to the duodenum.

See also

Liver & pancreas

The liver and the pancreas are both involved in digestion, but they also play a vital role in adjusting the chemical composition of your blood. Together, these two organs ensure that your blood is correctly balanced, so that your body can work at maximum efficiency.

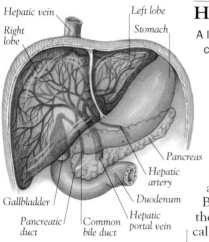

The liver
The spongy, wedge-shaped liver is richly supplied with blood. It has a large right lobe and a much smaller left lobe.

Liver

An organ that stores food substances and produces bile

The liver is the body's largest internal organ, and in an adult, weighs about 3 lb (1.4 kg). It is divided into two overall lobes, and fills most of the upper right-hand side of the abdomen ■. The liver carries out hundreds of different chemical reactions, and stores vital chemicals such as vitamins and glycogen ■ until they are needed. In digestion ■, its only function is the secretion of bile. Its main role is to collect blood from the alimentary canal ■ via the hepatic circulation ■, and adjust its chemical content before letting it flow around the rest of the body.

Hepatocyte

A liver cell that adjusts the composition of blood

The liver contains millions of hepatocytes, packed into six-sided columns called **lobules**. Hepatocytes are the liver's chemical "workshops." They are bathed in blood from the alimentary canal, and they absorb and release substances as the blood flows past.
Between sheets of hepatocytes there are blood-filled spaces, called **sinusoids**. These contain macrophages ■ called **Kupffer cells**, which engulf old red blood cells, bacteria, and debris by the process of phagocytosis ■. The main functions of hepatocytes are shown in the table opposite.

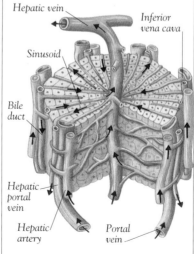

Anatomy of a liver lobule
This diagram shows the direction of the flow of blood and bile in a liver lobule.

Deamination

The removal of an amino group from an amino acid

If you eat lots of protein ■, your liver converts the excess amino acids ■ you absorb into glycogen, which it stores. However, before it can do this, the liver has to get rid of the nitrogen that the amino acids contain. It removes amino groups (NH_3) ■, and builds them into urea ■. The unwanted nitrogen is then excreted from the body by the kidneys ■.

Bile

A digestive fluid formed in the liver

Bile is a greenish yellow liquid made by the liver. It is about 97 percent water. Bile contains excretory substances, including cholesterol ■ and the pigment **bilirubin**, which is made when red blood cells ■ are broken down. This substance gives bile its color. Bile also contains **bile salts**, which are made in the liver and used in digestion. Bile flows into the duodenum ■, which is part of the small intestine ■. Here, the bile salts **emulsify** fats, which means that they turn them into small droplets that are easier to digest.

Gallbladder

A sac that stores bile

The gallbladder is a small, bag-like organ that is tucked underneath the liver. It is connected to the liver and the duodenum by tubes called **bile ducts**. The gallbladder collects bile from the liver, and concentrates it by removing most of its water. When semidigested food enters the duodenum, the muscular walls of the gallbladder pump bile into the small intestine. Together, the gallbladder and its ducts form the **biliary system**.

MAJOR FUNCTIONS OF THE LIVER

Function	Processes involved	Function	Processes involved
Blood sugar regulation	Absorbs excess glucose and stores it as glycogen; releases glucose if blood sugar level falls	Protein metabolism	Collects amino acids and uses them to make proteins; breaks down surplus amino acids through the process of deamination
Fat metabolism	Converts fats into a form that can be stored or broken down to release energy; also makes most of the body's cholesterol	Bile production	Forms bile, and makes the salts that are dissolved in it
Vitamin storage	Stores several vitamins, including vitamins A, D, and B_{12}	Detoxification	Removes poisonous chemicals from the blood, and breaks them down
Mineral storage	Stores iron and copper, two minerals needed to make hemoglobin	Hormone breakdown	Removes hormones from the blood and breaks them down

Background picture: micrograph showing section of liver

Pancreas

An organ that produces digestive fluids and regulates blood sugar levels

The pancreas is a long, slender organ that lies almost horizontally beneath the stomach. It forms part of both the digestive system and the endocrine system. The pancreatic cells that secrete enzymes used in digestion form exocrine glands ■ called **acini** (singular **acinus**). An acinus consists of a round cluster of cells that release pancreatic juice into the duodenum through a tube called the **pancreatic duct**. The pancreatic cells that produce hormones ■ are endocrine glands called **islets of Langerhans**. They produce the hormones insulin ■ and glucagon ■, which control the level of glucose in the blood. The islets of Langerhans release hormones directly into the bloodstream.

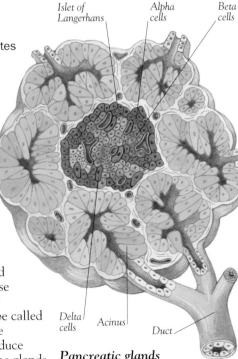

Islet of Langerhans · Alpha cells · Beta cells · Delta cells · Acinus · Duct

Pancreatic glands

Pancreatic endocrine glands consist of clusters of cells called islets of Langerhans. There are three main types of cell – alpha cells, beta cells, and delta cells. Surrounding these cells are groups of exocrine glands called acini. Acini secrete digestive enzymes into small ducts that lead to the pancreatic duct.

Pancreatic juice

A digestive fluid released by the pancreas

Pancreatic juice contains water, enzymes ■, and sodium bicarbonate. The enzymes include **pancreatic amylase**, which digests starch, **lipase**, which digests fats, and **trypsin**, which digests proteins. The sodium bicarbonate makes the juice slightly alkaline, so that it can neutralize the acid flowing from the stomach. Pancreatic juice is released in response to hormones and nervous impulses ■ produced as the stomach empties.

See also

Intestines

Your intestines make up about 80 percent of the total length of your digestive system. They are found in the lower part of your abdomen, and are coiled and folded to fit into this confined space.

Intestine

A tube that digests food and absorbs nutrients and water

The first part of the intestine is the small intestine. It carries out digestion ■ and absorption ■. The second part is the large intestine. This plays no part in digestion and its main function is to absorb water and salts. Both parts of the intestine are lined with a slippery mucous membrane ■. This membrane is constantly being worn away, but is replaced by rapid cell division.

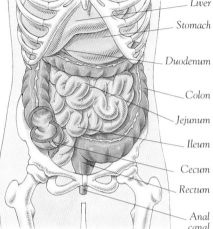

Small and large intestines
The intestines occupy much of the space in the central and lower abdomen.

Liver
Stomach
Duodenum
Colon
Jejunum
Ileum
Cecum
Rectum
Anal canal

Segmentation

The contraction of the intestinal wall to make closed compartments

Digested food moves through the intestines by waves of muscle contraction called peristalsis ■. Segmentation is a different kind of muscle contraction, in which the intestinal walls contract to form separate compartments. Segmentation ensures that the contents of the intestines are thoroughly mixed.

Small intestine

A part of the intestine that digests food and absorbs nutrients

The small intestine begins at the pyloric sphincter ■, just below the stomach, and ends at the cecum. It is divided into three parts – the duodenum, the jejunum, and the ileum. In an adult, the small intestine is about 22 ft (6.5 m) long, but just 1 inch (2.5 cm) across. The small intestine completes the process of digestion, and allows nutrients and water to be absorbed into the bloodstream. Its cells produce a variety of digestive enzymes ■. These include **peptidase**, which breaks down protein ■ fragments called peptides, **maltase**, which breaks down maltose ■, and **sucrase**, which breaks down sucrose ■. Children and most adults also produce lactase, which breaks down lactose ■.

Intestinal juice

A fluid in the small intestine

Intestinal juice is a watery liquid that contains partly digested food from the stomach. It carries these food substances to the walls of the small intestine. Here, they are digested and then absorbed by the surface cells of villi.

Villus

A microscopic projection in the lining of the small intestine

When seen through a microscope, the inner surface of the small intestine looks like a forest of tiny fingers or flaps. Each one is a villus (plural **villi**), and is up to 0.04 inch (1 mm) high. Villi contain a network of capillaries ■ and lymph vessels ■ called lacteals ■. Their surface cells have even smaller projections called **microvilli**. Together, the villi and microvilli provide the small intestine with a huge surface area for absorbing nutrients from digested foods into the blood and lymph ■.

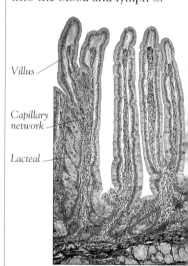

Villus
Capillary network
Lacteal

Villi in the small intestine
This light micrograph shows the finger-like villi that project from the lining of the small intestine. Villi and microvilli provide the small intestine with a surface area the size of a tennis court.

Duodenum

The first part of the small intestine

The duodenum is a C-shaped tube about 10 inches (25 cm) long. It receives chyme ■ from the stomach, digestive fluids from the pancreas ■, and bile ■ from the gallbladder ■. It also releases its own digestive enzymes. The lining of the duodenum has flaplike villi covered with alkaline mucus. This protects it from acids and enzymes.

Jejunum

The middle part of the small intestine

The jejunum is about 8 ft (2.5 m) long. It is tightly curved, and produces enzymes that complete the process of digestion.

The large intestine

The main function of the large intestine is to change liquid chyme from the small intestine into solid feces. Feces consist mainly of bacteria, dietary fiber, and dead cells from the lining of the alimentary canal.

Ileum

The last part of the small intestine

The ileum is about 13 ft (4 m) long. It is highly curved, and is lined with finger-shaped villi. The ileum plays only a small part in digestion. Its main function is to absorb nutrients from food that has been digested in the stomach, the duodenum, and the jejunum. The ileum ends at the **ileocecal sphincter**, which connects it to the caecum.

Large intestine

A part of the intestine that absorbs water and concentrates waste

The large intestine is a tube about 5 ft (1.5 m) long and 2.5 inches (6.5 cm) across. It does not carry out digestion, but it does absorb substances produced by bacteria in undigested waste, such as vitamin K. The large intestine also helps to maintain the body's fluid balance ■ by absorbing water. It has four parts – the cecum, the colon, the rectum, and the anal canal.

Cecum

The first part of the large intestine

The cecum is a short pouch. Attached to it is a small tube called the **appendix**. In the human body, both play little part in digestion.

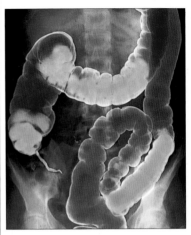

X-ray of the colon
This false color X-ray of the abdomen shows part of the large intestine.

Colon

The main part of the large intestine

The colon is a wide tube that travels up, across, and finally down the abdomen. It is divided into four parts – the **ascending colon**, the **transverse colon**, the **descending colon**, and the **sigmoid colon**. The colon absorbs about 90 percent of the water in undigested waste, and changes the waste from a liquid into solid masses called **feces**.

Rectum

The last part of the large intestine

The rectum is a short chamber that holds feces before they leave the body by the process of **defecation**. The **anal canal** leads from the rectum to an opening called the **anus**. The anus is kept closed by two sphincter muscles – the **internal anal sphincter** and the **external anal sphincter**.

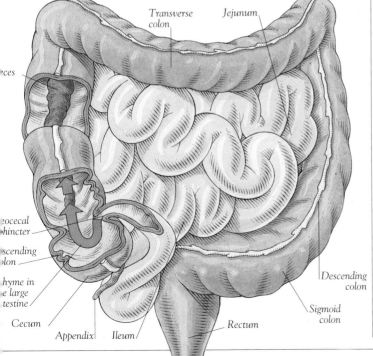

Transverse colon — Jejunum — Ileocecal sphincter — Ascending colon — Chyme in the large intestine — Cecum — Appendix — Ileum — Descending colon — Sigmoid colon — Rectum — Feces

Urinary system

Every hour, your kidneys remove up to 1.5 gallons (7 liters) of liquid from your blood. This liquid is filtered, and any useful substances are returned to your blood. The remaining liquid is flushed out of your body, carrying a variety of waste products that your cells need to get rid of.

Urinary system

A system that regulates the water and chemical content of the blood

All living cells produce chemical waste that is carried away by the blood ■. This waste is potentially poisonous, so it has to be removed, or excreted ■, from the blood before it has a chance to build up. The urinary system is made up by the kidneys, the bladder, the ureters, and the urethra. It disposes of nitrogenous waste ■, which is waste that contains nitrogen, in a liquid called urine. The urinary system also removes surplus water and salts from the blood. This ensures that its volume and osmotic pressure ■ remain within set limits.

Kidneys

Ureter

Bladder

See also

Antidiuretic hormone 79 • Blood 82
Blood pressure 89 • Excretion 77
Hormone 78 • Nitrogenous waste 77
Osmotic pressure 29 • Peristalsis 117
Pituitary gland 79
Plasma protein 83 • Sphincter 116

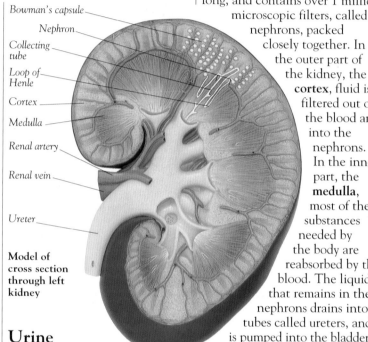

Bowman's capsule
Nephron
Collecting tube
Loop of Henle
Cortex
Medulla
Renal artery
Renal vein
Ureter

Model of cross section through left kidney

Urine

A fluid produced by the kidneys that carries away dissolved waste

Urine is about 95 percent water. The rest is made up mainly of dissolved salts, and a substance called **urea**, which contains nitrogen. Urine is formed by the kidneys, and the amount of urine they produce is regulated by hormones ■. If the blood becomes too concentrated, the pituitary gland ■ increases production of antidiuretic hormone (ADH) ■. This decreases the amount of urine produced and conserves water. If the blood becomes too dilute, the levels of ADH fall, and urine production increases.

Kidney

An excretory organ that eliminates waste and controls water content

The two kidneys are brown, bean-shaped organs that lie near the back of the abdomen. They remove waste products, salts, and water from the blood, and dispose of them as urine. Each kidney is up to 5 inches (12.5 cm) long, and contains over 1 million microscopic filters, called nephrons, packed closely together. In the outer part of the kidney, the **cortex**, fluid is filtered out of the blood and into the nephrons. In the inner part, the **medulla**, most of the substances needed by the body are reabsorbed by the blood. The liquid that remains in the nephrons drains into tubes called ureters, and is pumped into the bladder.

Nephron

A filtering unit in the kidney

A nephron has three main parts – a glomerulus, a Bowman's capsule, and a renal tubule (the word renal is used to describe anything connected with the kidneys). The Bowman's capsule collects fluid from the blood. As the fluid travels along the tubule useful substances such as glucose are reabsorbed by the blood, along with about 99 percent of the water in the tubule. By the time the fluid reaches the far end of the tubule it contains only those substances that the body needs to get rid of.

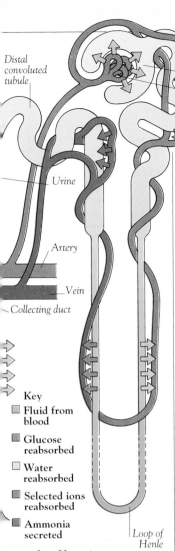

Distal convoluted tubule

Proximal convoluted tubule

Bowman's capsule

Glomerulus

Urine

Artery

Vein

Collecting duct

Key
- ■ **Fluid from blood**
- ■ **Glucose reabsorbed**
- □ **Water reabsorbed**
- ■ **Selected ions reabsorbed**
- ■ **Ammonia secreted**

Loop of Henle

omerular filtration
is diagram shows some of the stances that move through the lls of a nephron. The strength of the low colouring shows the changing ncentration of the waste fluid.

lomerulus

e microscopic cluster of pillaries at the head of a nephron

he word glomerulus means aped like a small ball." Inside glomerulus, blood pressure ■ ces fluid out of the capillaries, d into a nephron. The walls of glomerulus act like a filter. ey allow salts and water to ave the blood, but they block e path of blood cells and most asma proteins ■.

Bowman's capsule

The cuplike structure surrounding the glomerulus

Each nephron has a Bowman's capsule, or **glomerular capsule**, in the outer layer or cortex of the kidney. The capsule forms the closed end of a nephron's tube. The wall of the capsule is "leaky," or permeable, so that it can collect fluid that is forced out of the glomerulus. The fluid then enters the renal tubule.

Renal tubule

A tube that allows substances to be reabsorbed by the blood

A renal tubule is a microscopic tube that makes up the largest part of each nephron. When fluid enters the tubule from the Bowman's capsule, it contains useful substances such as glucose, together with salts and a large amount of water. As it flows along the tubule, all the glucose is reabsorbed by the blood, along with most of the water, and some of the salts. Each nephron has a single tubule, divided into three parts. The **proximal convoluted tubule** and the **distal convoluted tubule** are twisted and relatively thick. The central section, the **loop of Henle**, is long and thin.

Ureter

A tube that carries urine from the kidney to the bladder

At the end of a renal tubule is a tube called the **collecting duct**, into which urine drains. The ducts in each kidney lead to a ureter, which empties into the bladder. The two ureters have muscular walls, and move urine by peristalsis ■.

Bladder

A hollow organ that stores urine

The bladder is an elastic bag at the base of the abdomen. It receives urine from the two ureters, which open into its lower surface. Urine leaves the bladder through a tube called the **urethra**. Normally, the urethra is closed off by two circles of muscle, or sphincters ■. During **urination**, the sphincters relax so that urine can leave the bladder.

Dialysis

An artificial way of removing waste and surplus fluid from blood

If someone suffers from **renal failure**, their kidneys do not work normally, and waste products and fluid gradually build up inside their blood. Without treatment, this condition can be very dangerous. Dialysis is a way of treating this disorder. In the most common method, called **hemodialysis**, the patient's blood is passed through a machine called an **artificial kidney**, or **dialyzer**. This allows waste substances and fluid to pass out of the blood, before the blood is then returned to the body. A single dialysis session can take several hours, and has to be repeated every few days.

Dialysis
This patient is undergoing hemodialysis. The dialysis machine replaces the function of the kidneys.

Reproductive system

Reproduction ensures the continuation of the human species. The reproductive system differs in men and women, and it does not start to function until the body is approaching maturity.

Reproductive system

A body system that produces offspring

Humans produce children through **sexual reproduction**. This involves special sex cells produced by a man and a woman. These are brought together during sexual intercourse ■ to form new individuals. The reproductive organs of men and women are called **genitals**. During pregnancy ■, a woman's reproductive system provides a protective environment for a developing baby.

Sex cell

A cell that is involved in sexual reproduction

Sex cells, or **gametes**, are formed by a process called **gametogenesis**. This involves a special kind of cell division ■ called meiosis ■, which results in sex cells being haploid ■, or having half the number of chromosomes ■ found in other body cells. During sexual reproduction, a male sex cell, or sperm, combines with a female sex cell, or ovum, and a single new cell is formed.

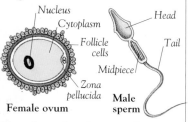

Female and male sex cells
In real life, a female ovum is far larger than a male sperm. The ovum is 50 times as wide as the head of the sperm.

Sperm

A male sex cell

A sperm cell, or **spermatozoon** (plural **spermatozoa**), is about 0.002 inch (0.05 mm) long. It has an oval head that contains a nucleus ■, a cylindrical midpiece that contains mitochondria ■, and a slender tail. The mitochondria release energy that enables the tail to lash from side to side, and so move the sperm cell along. In an adult male, more than 250 million sperm cells are made every day, by a process called **spermatogenesis**.

Testis

The reproductive organ in males that produces sperm

The two **testes** (plural of testis) are located outside the body, in a pouch called the **scrotum**. Here, they stay cooler than the rest of the body, which is essential for sperm production. Each testis contains a set of coiled tubes, or **seminiferous tubules**, which make sperm cells. The sperms move into the **epididymis**, a large coiled tube, where they mature. The epididymis leads into a muscular tube – the **vas deferens**. This carries sperm cells to the penis. The testes also produce male sex hormones ■ to prepare a man's body for reproduction.

Prostate gland

The gland in males that helps to form the fluid in which sperms sw■

Sperm cells swim in a liquid called **seminal fluid**, or **semen** (a word that means seed). The fluid stops the sperm from dryi■ out, and it provides them with sugary fuel for their journey inside a woman's body. The prostate gland, located below t■ bladder ■, produces part of this liquid. The rest is made by two glands called the **seminal vesicl■**

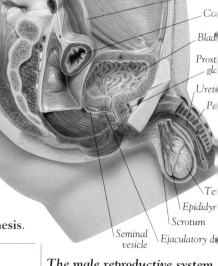

Ure■
Co■
Blad■
Prost■
gl■
Uret■
Pe■
Te■
Epididyr■
Scrotum
Seminal vesicle
Ejaculatory d■

The male reproductive system
The main organs of the male reproduct■ system are located outside the body.

Penis

The organ in males that transfers sperm into the female's body

The penis contains a central channel, called the urethra ■, which is surrounded by column■ of spongy tissue. Before sexual intercourse takes place, the spongy tissue fills with blood, and the penis becomes erect. This change in shape allows it t■ be inserted into the vagina, so that sperm cells can be transferre■ from the male to the female. Sperm cells leave the man's bod■ in a process called ejaculation ■

Ovum

the female sex cell

A single ovum (plural **ova**), or **egg cell**, is about 0.04 inch (0.1 mm) across. It contains a single set of chromosomes and is almost round. It is surrounded by a jelly-like film called the **zona pellucida**, which is itself surrounded by a layer of small follicle ■ cells. Ova are formed from cells called **oocytes**, through a process called **oogenesis**. Oogenesis is completed before birth. Many years later, during adulthood, the ova are released one by one in a process called ovulation ■. If an ovum is fertilized ■ by a sperm, it develops into an embryo ■.

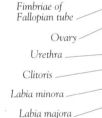

Fallopian tube
Fimbriae of Fallopian tube
Ovary
Urethra
Clitoris
Labia minora
Labia majora

Ovary

the reproductive organ in females that produces ova

Ovaries are two paired glands ■, about 1.2 inches (3 cm) long, that produce a woman's ova. They are situated on either side of the uterus, and each one is partly surrounded by the funnel of a Fallopian tube. At birth, the ovaries contain about 1 million immature ova. This is vastly more than is needed, but no more are formed in later life. From puberty ■ onwards, the ovaries release one mature ova approximately every month from sacs called follicles. The ovaries also produce female sex hormones to prepare the body for pregnancy.

Uterus

The organ in females in which a baby develops

The uterus, or **womb**, is a hollow chamber with muscular walls. The upper part has two openings that lead into the Fallopian tubes. The lower end has one narrow opening, called the **cervix**, that leads into the vagina. When a woman is not pregnant, the uterus is folded almost flat and is about 3.1 inches (8 cm) long. During pregnancy, it becomes far larger.

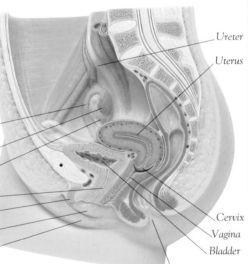

Ureter
Uterus
Cervix
Vagina
Bladder
Anus

The female reproductive system
Most female reproductive organs are located inside the body.

Fallopian tube

The channel in females that carries ova from an ovary to the uterus

The two Fallopian tubes connect the ovaries with the uterus. Each one is about 3 inches (7.5 cm) long, and has a funnel-like mouth fringed with projections called **fimbriae** (singular **fimbria**). When an ovary releases an ovum, cilia ■ on the fimbriae sweep the ovum along the tube. Waves of muscle contraction pass down the tube, and help to push the ovum into the uterus. Fertilization usually takes place in the Fallopian tube.

Antony van Leeuwenhoek

Dutch microscopist (1632–1723)

Leeuwenhoek made lenses for use in microscopes. He made many discoveries about cells and microorganisms, and was one of the first people to see sperm. Some scientists thought that sperm contained tiny preformed humans, called **homunculi**. Others believed that preformed humans existed in egg cells. As microscopes increased in power, homunculi were never found, so both sides were proved to be wrong.

Vagina

The muscular tube leading from the entrance of the female reproductive tract to the uterus

During sexual intercourse, sperm enter a woman's body through the vagina. During birth ■, a baby passes through the vagina, which stretches wide to let the baby out. The entrance to the vagina is surrounded by two paired folds of tissue called the **labia majora** and between these, the **labia minora**. The **clitoris**, an organ at the top of the labia minora, becomes erect during sexual stimulation. Together, the labia and the clitoris form the external female genitals.

See also

Reproductive cycle

A woman's reproductive cycle prepares her body to nurture new life. In a series of closely linked changes, it triggers the release of an egg cell and creates the right environment in which the egg can develop.

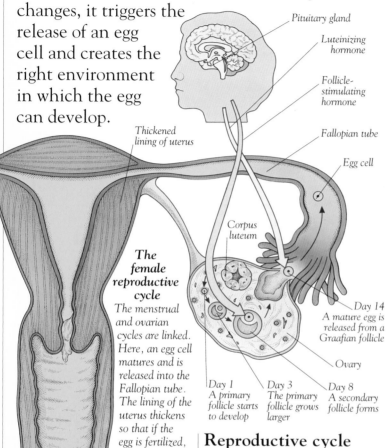

Pituitary gland

Luteinizing hormone

Follicle-stimulating hormone

Fallopian tube

Egg cell

Thickened lining of uterus

Corpus luteum

The female reproductive cycle
The menstrual and ovarian cycles are linked. Here, an egg cell matures and is released into the Fallopian tube. The lining of the uterus thickens so that if the egg is fertilized, it can implant itself in the uterine wall.

Vagina

Day 14
A mature egg is released from a Graafian follicle

Ovary

Day 1
A primary follicle starts to develop

Day 3
The primary follicle grows larger

Day 8
A secondary follicle forms

See also

Ovarian cycle

A cycle leading to the release of an ovum

Once a girl reaches puberty, her ovaries ■ begin to release egg cell or ova, at a steady rate – usually one every 28 days. The ovarian cycle starts when the pituitary gland ■ releases follicle-stimulating hormone ■. This triggers the development of follicles in the ovaries. A second pituitary hormone, called luteinizing hormone ■, then stimulates ovulation. If fertilization does not occur, the cycle begins again.

Follicle

A ball-shaped cluster of cells containing a single ovum

Ova are surrounded by cells that make up structures called follicles. Follicles nourish ova and protect them from damage. At the beginning of each ovarian cycle, about 24 follicles start to grow. Each one changes from a **primary follicle**, which is a solid ball of cells, into a **secondary follicle**, which contains a fluid-filled space. Eventually, one of the secondary follicles becomes so large that it forms a blister-like swelling, called a **Graafian follicle**, under the surface of the ovary. At the moment of ovulation, the Graafian follicle bursts. Its egg cell is swept into the Fallopian tube ■ and toward the uterus ■.

Reproductive cycle

A repeated sequence of changes in the reproductive system

From puberty, male sex cells ■ are produced all the time at a steady rate. If they are not used, they are broken down. The female reproductive system ■ works in a different way. A lifetime's supply of sex cells is present at birth, so no more have to be made. From puberty ■ onwards the sex cells mature in monthly batches. Each time this happens, a single ovum ■ is released so that fertilization ■ can occur. The female reproductive cycle has two interlinked parts – the ovarian cycle and the menstrual cycle.

Release of an egg cell
As the Graafian follicle swells in size, it bursts and releases its egg cell.

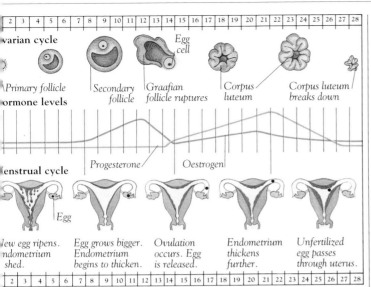

Ovarian cycle

Primary follicle — Secondary follicle — Graafian follicle ruptures — Egg cell — Corpus luteum — Corpus luteum breaks down

Hormone levels

Progesterone / Oestrogen

Menstrual cycle

Egg

New egg ripens. Endometrium shed. — Egg grows bigger. Endometrium begins to thicken. — Ovulation occurs. Egg is released. — Endometrium thickens further. — Unfertilized egg passes through uterus.

The ovarian cycle, ovulation, and the menstrual cycle

As an egg cell matures in an ovary, the ovary produces estrogen. This thickens the lining of the uterus in preparation for a fertilized egg implanting itself. The corpus luteum releases progesterone, which increases the blood supply to the endometrium. If fertilization does not occur, the uterine lining is shed 14 days after ovulation and the cycle begins again.

Ovulation

The release of an ovum

A woman's two ovaries usually alternate the release of one egg cell, or ovum. This means that each ovary releases an ovum about once every two months.

Corpus luteum

A structure formed by a follicle that has shed its ovum

Corpus luteum means "yellow body." It develops from a Graafian follicle after ovulation. If fertilization occurs, the corpus luteum produces the hormone progesterone ■. This helps to prepare the lining of the uterus for implantation ■ and pregnancy ■. If fertilization does not occur, the corpus luteum starts to break down and disintegrate about 10 days after ovulation.

Menstrual cycle

A cycle of changes in the lining of the uterus

The menstrual cycle lasts for about 28 days. It prepares the uterine lining, or **endometrium**, to receive an egg cell, but the lining breaks down if the egg does not implant. The cycle starts when primary follicles release oestrogen ■, making the endometrium thicken. After ovulation, progesterone from the corpus luteum increases the blood supply to the endometrium. If an egg implants, the cycle stops until after birth ■. If implantation does not occur, the lining and the egg are shed, and the cycle starts once again.

Menstruation

The breakdown and shedding of the lining of the uterus

During menstruation, the thickened lining of the uterus falls away, and its cells and blood flow out of the body through the vagina ■. Usually, the amount of blood lost is small and is quickly replaced. Menstruation normally lasts for about 5–7 days. This part of the menstrual cycle is commonly known as a **period**.

Menarche

The point at which the menstrual cycle starts for the first time

Most girls have their first period at about the age of 12 years. This is called menarche (pronounced men-arky). From this moment onward, a girl's reproductive system starts to function fully, although her body has not yet reached its final adult form.

Menopause

The point at which the menstrual cycle comes to an end

As a woman enters middle age, her ovarian and menstrual cycles become less frequent. Eventually, between the ages of about 45 and 55 years, they come to a halt, and the reproductive system no longer functions. Menopause is caused by changes in the levels of sex hormones ■ present in the body.

Regnier de Graaf

Dutch physician (1641–73)

In 1672, Regnier de Graaf published a book that described the female reproductive system in detail for the first time. His skill as an anatomist enabled him to explain the workings of the reproductive organs, but he did make one mistake. He thought that the swellings that he could see in ovaries were egg cells, or ova. What he actually saw were follicles, each with an egg cell hidden away inside. Despite this, his work was an important advance, and his name still survives in the term "Graafian follicle."

Genes & chromosomes

Unless you have an identical twin, you will have a unique set of genes. Genes are separate instructions in the body's chemical plan. They control thousands of different characteristics, from your height to the color of your hair.

Gene

A basic unit of heredity that carries the instructions needed to make a particular protein

Genes are instructions that are stored in the DNA ▤ inside a cell's nucleus ▤. Each gene tells a cell how to make a particular protein ▤. Proteins play a vital part in the life of cells, so by controlling the production of proteins, genes control the way that cells work. A normal cell contains about 100,000 genes, arranged in two matching sets of chromosomes. There are usually two alternative forms of each gene, called alleles ▤, one from the mother and one from the father. Normally, only one allele from each pair is put into action. Genes are passed on during sexual reproduction ▤. By this process, called heredity ▤, inherited characteristics ▤ pass from one generation to the next.

Genetic code

The chemical code used by DNA

Genes are made up by a long sequence of chemicals called bases ▤, arranged in a precise but varying order. Using the genetic code, a cell can convert the sequence of bases into a sequence of amino acids ▤, to make a protein. The genetic code uses "words" called **codons**, which are each three bases long. A codon instructs the cell to select a particular amino acid, and to add it to a growing protein molecule.

Protein synthesis

The manufacture of a protein

To make a protein, a cell has to follow instructions from a gene. This process is called protein synthesis. In the first step of protein synthesis, called **transcription**, a cell makes a copy of a gene, using a nucleic acid ▤ called messenger RNA ▤, or mRNA. The second step is called **translation**. It uses a nucleic acid called transfer RNA ▤, or tRNA, and chemical clusters called ribosomes ▤ to assemble amino acids in the right sequence to build the protein.

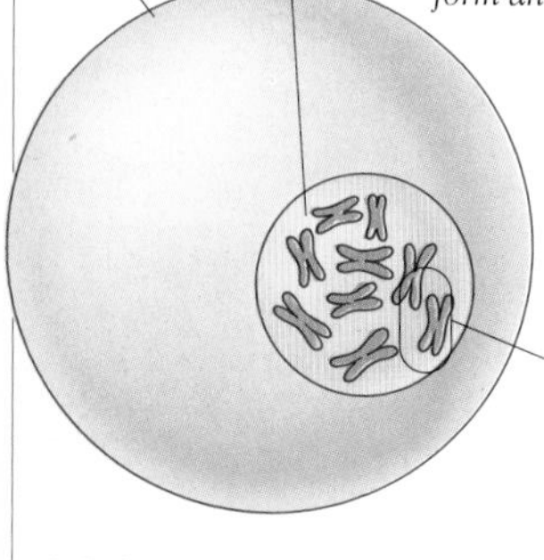

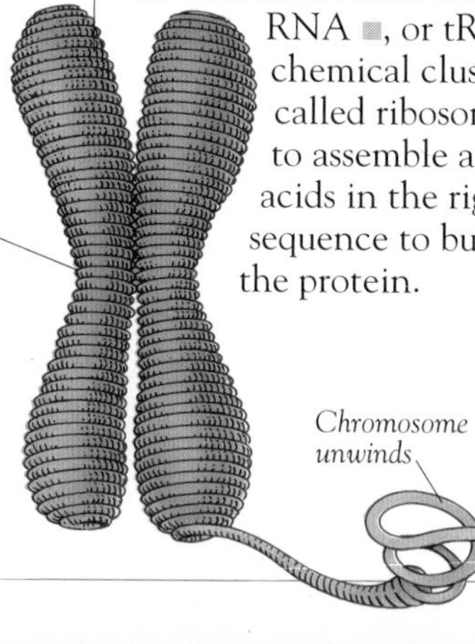

Cell   Nucleus
Chromatids in a condensed chromosome form an X-shape

Making a protein
This illustration shows how cells use the instructions in a gene to make a protein.

Chromosome unwinds

Genotype

The genetic makeup of a cell, or of the body as a whole

A person's genotype consists of the entire collection of alleles that they inherit from their parents. Only some of these allel⟨ are put into action, or **expresse⟨** Others remain **unexpressed**, which means that they play no part in shaping or controlling th body. Both kinds of alleles can ⟨ passed on in sexual reproductio⟨ and an allele that is unexpresse⟨ in a parent may become expressed in their children.

Phenotype

The visible characteristics produced by a genotype

A person's phenotype is all of their physical characteristics. Th⟨ phenotype is partly programme⟨ by genes, but it is also shaped b⟨ external factors, such as exercis⟨ and diet ▤. These factors work with the genetic plan to produc⟨ the body's shape.

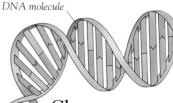

DNA molecule

Chromosome

A structure that contain⟨ part of a cell's DNA

Chromosomes are threadlike packages of DNA. They are stored in a cell's nucleus, and are normally visible only wher⟨ they wind up, or **condense⟨** just before a cell divides ▤. At this stage, a chromosome has two identical arms, or **chromatids**. Humans have 46 chromosomes in each cell. This number is called the **chromosome number**.

Diploid cell

A cell that has a double set of chromosomes

Almost all the cells in the body have two sets of chromosomes. Each one contains copies of chromosomes that originally came from two different sources – one set from the father, and one set from the mother.

Haploid cell

A cell with a single set of chromosomes

Haploid cells, or sex cells ■, are formed by the special kind of cell division called meiosis ■. Haploid cells are used in sexual reproduction, and contain only 23 chromosomes instead of the usual 46. During fertilization ■, a male and female sex cell combine to produce a new diploid cell.

DNA molecule unwinds

Messenger RNA forms

DNA rewinds

Messenger RNA copies the DNA's information

1 *Transcription begins when part of a DNA molecule unwinds. A single-stranded molecule of messenger RNA is then built up piece by piece. The messenger RNA then leaves the cell's nucleus and enters its cytoplasm, where it latches onto a ribosome.*

Homologous chromosome

One of a matching pair of chromosomes in a diploid cell

A diploid cell contains two sets of chromosomes. Each set originally comes from a different parent. The chromosomes form matching pairs called homologous chromosomes, or **homologues**. Although they look identical, the chromosomes in each pair often contain different versions of the same genes. These versions are known as alleles, and they can have different effects.

Sex chromosome

A chromosome that determines sex

There is an important difference between male and female cells. In women, diploid cells always contain a pair of matching **X chromosomes**. In men, there is only one X chromosome, which is paired with a much smaller **Y chromosome**. The combination of X and Y chromosomes determines a person's sex. Because of the way meiosis works, most people have one of two combinations – XX or XY. Other combinations are rare. **Autosomes** are chromosomes that do not determine sex.

Outside nucleus

Inside nucleus

Ribosome

Genome

The complete set of genes found in the body

The human genome consists of all the genes that we possess. In recent years, scientists have been carrying out a major research program, called the **Human Genome Project**, which aims to pinpoint all our genes, and find out what each one does.

2 *During translation, a strand of messenger RNA (mRNA) attaches itself to a ribosome. The ribosome reads the mRNA's bases one codon at a time. With the help of transfer RNA, it builds up a chain of amino acids to form a protein molecule. Once translation is complete, the finished protein can be used.*

Amino acid chain builds up

Amino acid

Transfer RNA

Strand of messenger RNA

Karyotype

A complete set of chromosomes inside a cell

In a karyotype, the chromosomes inside a cell are photographed and arranged into homologous pairs. This can be done by rearranging the photograph by hand or with a computer.

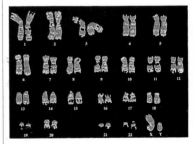

Male karyotype
This light micrograph shows a male karyotype. The male sex chromosomes, labelled "XY," are shown bottom right.

Heredity

Every child receives genes from both its parents. But instead of pooling their effect, genes work in a more complicated way. Sometimes a gene will be expressed, but it may remain inactive until it is passed to the next generation.

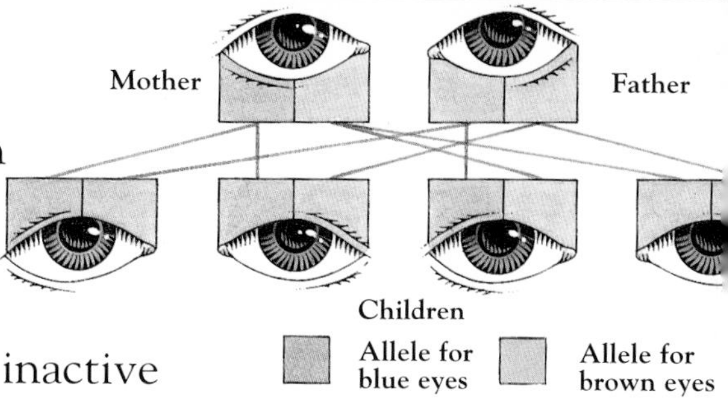

Mother

Father

Children

Allele for blue eyes

Allele for brown eyes

Dominant and recessive alleles
The allele for brown eyes is dominant and the allele for blue eyes is recessive. This diagram shows what happens whe the two alleles are passed on.

Heredity

The transmission of characteristics controlled by genes

When people have children, they pass on a set of their genes ▪ to their offspring. These genes produce **inherited characteristics**, which are all the features that genes control. Some inherited characteristics, such as blood groups ▪, are controlled by single genes, but many others are controlled by several genes working together. Genes are passed on according to precise mathematical patterns, and they interact with each other in many different ways. The study of these patterns and interactions is called **genetics**.

Allele

One of two or more forms of the same gene

In most of the body's cells, there are two sets of matching or homologous chromosomes ▪. The chromosomes in each pair carry alternative versions of the same genes, and these are called alleles, or **allelomorphic genes**. The alleles for each gene are normally in the same position, or **locus**, on matching chromosomes. They control the same characteristic, but often only one of the alleles is put into effect. In the human species as a whole, many genes have dozens of alleles. However, only two alleles for each gene can exist in a single cell.

Dominant allele

An allele that usually shows itself in the body's phenotype

When two different alleles exist in the same cell, only one usuall produces any effect. This is known as the dominant allele, o the **dominant gene**. The other allele is hidden, or masked.

Recessive allele

An allele that is usually masked by a dominant allele

When a recessive allele, or **recessive gene**, exists in the same cell as a dominant allele, its effect is masked, and it does not produce any outward effect. However, it still forms part of the cell's genotype ▪. In later generations, it may be partnered by another recessive allele. If thi happens, it will be expressed, an will produce a noticeable effect.

Codominant allele

An allele that always has some effect

Some genes have alleles that always produce an effect – none are truly dominant or recessive. An example is the gene that determines ABO human blood grouping. This has three alleles. The alleles that produce blood groups A and B are codominant.

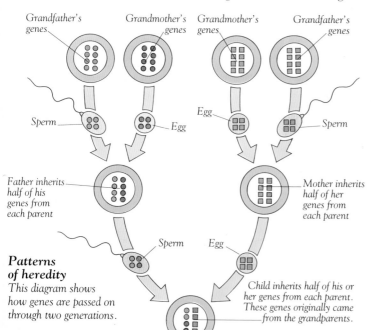

Grandfather's genes

Grandmother's genes

Grandmother's genes

Grandfather's genes

Sperm

Egg

Egg

Sperm

Father inherits half of his genes from each parent

Mother inherits half of her genes from each parent

Sperm

Egg

Patterns of heredity
This diagram shows how genes are passed on through two generations.

Child inherits half of his or her genes from each parent. These genes originally came from the grandparents.

Hemophilia

This diagram shows how x-linked disorders, such as hemophilia, can be passed down through generations.

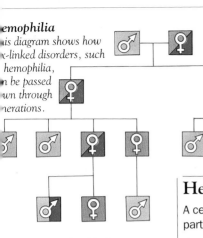

Key

- Unaffected female
- Unaffected male
- Female carrier
- Affected male

Sex-linked allele

An allele that is carried by a sex chromosome

Sex chromosomes do not only determine a person's sex. The X chromosome carries many alleles that control other body characteristics, and it is likely that the much smaller Y chromosome carries some too. Many sex-linked alleles occur only on the X chromosome. Because women have two X chromosomes, the effect of a recessive allele is often masked by a dominant allele. However, men have only one X chromosome, so the recessive allele is always expressed. Hemophilia and color blindness are two genetic disorders that are brought about by these kind of alleles. Such conditions are widespread in men, but rare in women.

Homozygous cell

A cell with identical alleles for a particular characteristic

A cell that contains two copies of the same allele is said to be homozygous for the particular characteristic the allele produces. The alleles are either both dominant, or both recessive. The characteristic that they produce will always be put into effect, or expressed, because neither allele masks the other.

Heterozygous cell

A cell with different alleles for a particular characteristic

A cell that is heterozygous for a particular characteristic contains two different alleles – one dominant, and one recessive. Normally, only the dominant allele will have any effect.

Mutation

A change in a cell's genetic material

During sexual reproduction, genes are rearranged to combine inherited characteristics in new ways. Genes can also be changed to produce new characteristics, through mutations. Mutations are chemical accidents that affect either a short piece of DNA or a whole chromosome. They occur naturally, but their rate increases if the body is exposed to harmful chemicals or radiation. In a normal body cell, a mutation is passed on when the cell divides, but lasts only one lifetime. In a sex cell, a mutation can be passed from one generation to another.

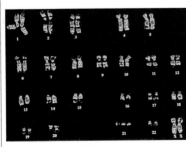

Genetic mutation
Down syndrome is the result of an extra chromosome 21, as shown here.

Genetic disorder

A disorder triggered by genes

The body's genes normally contain all the instructions that are needed to build it, and to make it work. However, they can also contain instructions that cause things to go wrong. Some genetic disorders, such as Down syndrome, occur because a person has an unusual number of chromosomes. Other disorders are triggered by single genes, or by small groups of genes. Some of these disorders are so severe that they usually prove fatal before birth. Others become apparent only much later in life.

Genetic engineering

The artificial alteration of a cell's genotype

In recent years, scientists have discovered ways to take genes from one cell, and to make them work when inserted into another cell. At present, this technique is used mainly to insert human genes into other organisms. For example, the gene that codes for the human hormone insulin has been successfully transferred into bacteria. The bacteria are then grown in laboratories, and the insulin they make is "harvested," so that it can be used to treat people with diabetes. In the future, genetic engineering may be used to alter human cells, so that dangerous genetic disorders can be treated.

Beginning of life

Every person on our planet begins life as a single fertilized egg. The process of fertilization brings together cells from two people, to create a new and unique human being.

Sexual intercourse

The process that brings male and female sex cells together

Sexual intercourse is the usual way humans reproduce. During sexual intercourse, or **copulation**, a man places his penis ▪ inside a woman's vagina ▪. Their movements trigger a reflex ▪ action, which makes the man release seminal fluid ▪, or **ejaculate**, into the woman's body. Sexual intercourse can produce enjoyable sensations; the desire to have sex strongly affects people's behavior ▪.

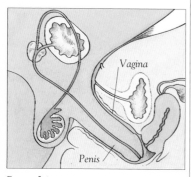

Sexual intercourse
A man places his penis inside a woman's vagina and releases sperm into her body.

See also

Fertilization

The union of a sperm and an ovum

After sexual intercourse, sperm ▪ travel through the vagina and uterus ▪ until they meet an egg cell, or ovum ▪. This usually occurs in a Fallopian tube ▪. The sperm release enzymes ▪ to dissolve the small follicle ▪ cells around the egg. Although thousands of sperm may have reached the egg only one will fertilize it. Eventually, one of the sperm cells breaks through the film around the egg called the zona pellucida ▪. The nucleus from the sperm fuses with the nucleus in the egg, and a new cell, or zygote, is formed. This is the moment of fertilization. Once fertilized, the egg undergoes chemical changes to prevent other sperm from attaching themselves to the egg. For fertilization to be successful, sperm cells have to reach the ovum within 24 hours of it being released. Sperm can survive for up to 72 hours inside the female body, so an ovum can be fertilized if sexual intercourse occurs before ovulation. Sexual intercourse does not always result in fertilization.

Fertilization
This sperm is just about to break through the zona pellucida surrounding the egg, so that fertilization can occur.

Egg being fertilized by sperm   Zygote   Fallopian tube

Egg released from ovary

Zygote

A cell produced when a sperm fertilizes an ovum

A zygote contains one set of chromosomes ▪ from each parent. The genes ▪ in these chromosomes contain all the instructions needed to produce a new human being, and to make its body work. Two sex chromosomes ▪, one from the mother and one from the father control the sex of a child. The sex is fixed at the zygote stage.

Cleavage

The division of a fertilized egg into many cells

Once an egg cell, or ovum, has been fertilized it divides in two. This process is repeated time and time again, so that many cells are formed. To begin with, the new cells, called **blastomeres**, get smaller as they divide so no growth ▪ occurs. But once the cluster of cells has implanted itself in the lining of the uterus, rapid growth begins.

Morula

A solid ball of cells formed from a fertilized egg

As the zygote divides, it forms a berrylike cluster of cells called a morula. The morula is still surrounded by the zona pellucida, which it inherits from the fertilized egg. This stage of development lasts for 3 days after fertilization. As cleavage continues, the morula travels down the Fallopian tube and towards the uterus.

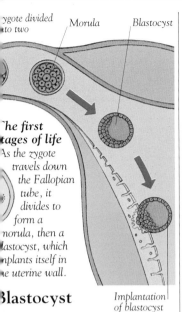

Zygote divided into two — Morula — Blastocyst

The first stages of life
As the zygote travels down the Fallopian tube, it divides to form a morula, then a blastocyst, which implants itself in the uterine wall.

Implantation of blastocyst

Blastocyst

A hollow ball of cells formed from a fertilized egg

As the morula nears the uterus it becomes hollow and is known as a blastocyst. This contains a fluid-filled space called the **blastocoele**, and has an outer layer of cells called the **trophoblast**. Initially, the blastocyst's cells are dispersed evenly around the fluid, but gradually a mass of cells forms on one region of the inner wall. The blastocyst enters the uterus about 5 days after fertilization, and the zona pellucida disintegrates. A day later, the blastocyst implants itself in the wall of the uterus.

Implantation

The attachment of a developing egg to the lining of the uterus

During implantation, the outer trophoblast cells of the blastocyst produce enzymes that dissolve some of the mother's cells in the uterus. Eventually, the blastocyst burrows right into the lining of the uterus, and becomes covered by the mother's cells. The inner cells of the blastocyst become the embryo; the outer cells form part of the placenta ■.

Conception

Fertilization followed by implantation

Conception covers the period from fertilization to the successful implantation of the blastocyst in the wall of the mother's uterus. At this stage, the mother is said to have conceived. Conception is not the same as fertilization, because a fertilized egg does not always manage to implant itself. Without implantation, it cannot develop.

Embryo

An unborn child in the first eight weeks of development

An embryo develops from the cell mass inside a blastocyst. Initially, it obtains all its nutrients from the mother's cells digested during implantation. The embryo continues to obtain most of its nutrients in this way for about 5–6 weeks, until an organ called the placenta is fully functioning. The embryo grows and develops very quickly, and by the end of 8 weeks, all the body's organs have begun to form. The amnion ■ and the chorion ■ also form early on to surround and protect the growing embryo.

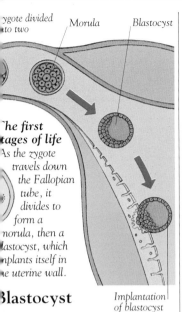

Five-week-old embryo
This embryo is floating in the fluid contained in the amniotic sac. Its arms and legs have started to form.

Fetus

An unborn child from the ninth week of development to birth

During the fetal stage, organs that have begun to appear in the embryo grow and develop rapidly. The fetus is initially about 1.2 inches (3 cm) long, and weighs about 0.035 oz (1 g). By the time it is ready to be born, it is about 20 inches (50 cm) long, and 3,000 times heavier.

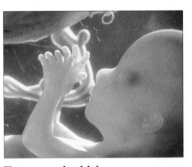

Four-month-old fetus
The head, arms, and hands are now very well developed. Blood vessels can be seen under the surface of the skin.

Differentiation

The specialization of cells as the body develops

As an embryo develops, its cells do not simply multiply. They also change, or **differentiate**, to adapt to particular tasks. Early in the life of an embryo, its cells form three layers, called the **ectoderm** (outer skin), **mesoderm** (middle skin), and **endoderm** (inside skin). Each of these layers develops to produce particular tissues ■ in the body. The ectoderm produces nerves, skin, hair, fingernails, and parts of some sense organs. The mesoderm forms bones, muscles, blood, and also sex cells ■. The endoderm forms epithelial tissue ■, which lines the alimentary canal ■, lungs, and other hollow organs.

Pregnancy

Over the course of about nine months, a single fertilized egg becomes transformed into a complete human being. During this process, the original cell divides many times over, to produce more than 10,000 billion new cells.

Pregnancy

The period from conception to birth

Pregnancy begins if fertilization ■ is followed by the implantation ■ of a growing egg, or blastocyst ■, in the uterus. The egg grows into an embryo ■ and then a fetus ■, using nutrients supplied by the mother. When development is complete, the baby is pushed out of the mother's body during birth ■. Pregnancy lasts for about 40 weeks, but 10 percent of pregnancies result in a premature ■ baby, which is born earlier. During pregnancy, the mother's body adapts to the fetus growing inside it, and prepares for the period after birth when the young baby is fed on milk ■.

Pregnant woman
A woman's abdomen swells gradually in pregnancy.

Placenta

An organ that links the blood supply of a mother and her baby

The placenta is a flat, sponge-like organ that develops from the growing embryo and the mother. It is attached to the lining of the uterus ■, and is connected to the embryo or fetus by the umbilical cord. The placenta works as a life support system. It brings the blood supply of mother and baby close together. Oxygen and nutrients diffuse ■ into the baby's blood from the mother's, and waste products from the baby diffuse in the other direction. The placenta also produces hormones ■. These help the mother's body adapt to pregnancy, and prepare her to produce milk.

Umbilical cord

A cord that connects the fetus to the placenta

The umbilical cord carries nutrients from the placenta to the embryo or fetus, and waste products in the other direction. It contains arteries ■ and veins ■ and is up to 24 inches (60 cm) long. After birth, the umbilical cord is cut, leaving a scar called the **navel**, or **umbilicus**. Because the cord does not contain any nerves, cutting it is painless.

Amnion

A fluid-filled membrane surrounding a developing embryo or fetus

The amnion is like a transparent balloon. It surrounds the embryo and later the fetus, but it becomes separated from it by a space containing liquid called **amniotic fluid**. The embryo or fetus floats in the fluid, which acts as a shock absorber and protects it from sudden jolts. During pregnancy, a small amount of amniotic fluid is sometimes extracted and tested, in a procedure called **amniocentesis**. By looking at cells in the fluid, doctors can give an early warning of genetic disorders ■ in the embryo.

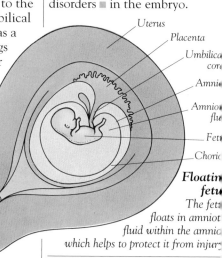

Uterus

Placenta

Umbilical cord

Amnion

Amniotic fluid

Fetus

Chorion

Floating fetus
The fetus floats in amniotic fluid within the amnion, which helps to protect it from injury.

Chorion

The outermost membrane surrounding the embryo or fetus

The chorion plays an important part in nourishing a growing fetus. Its finger-like projections, called **chorionic villi**, make up part of the placenta. The villi bring the blood of the mother and the fetus close together, so that substances can pass from one to the other. **Chorionic villus sampling** involves extracting chorion cells and examining them for any genetic disorders in the fetus. It can be performed as early as eight weeks of pregnancy.

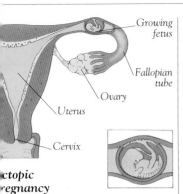

Growing fetus

Fallopian tube

Ovary

Uterus

Cervix

ctopic regnancy
his diagram shows e typical location of n ectopic pregnancy.

Fetus growing in a Fallopian tube

ctopic pregnancy

regnancy involving implantation utside the uterus

he word ectopic means out of place." In an ectopic regnancy, an egg implants n a Fallopian tube ▪, or nore rarely, in another art of the reproductive »stem. The embryo cannot evelop successfully outside the terus, and without surgery the nother may die. About 0.5 per-ent of pregnancies are ectopic.

Multiple pregnancy

pregnancy involving more than ne fetus

Iultiple pregnancies occur when nore than one ovum ▪ is released nd then fertilized, or when a ngle fertilized ovum divides. The nost common form of multiple regnancy involves two babies, r **twins**. Three babies, called riplets, and four babies, called uadruplets, are much rarer, and re usually a result of the mother aking fertility drugs to improve er chances of onception ▪. Multiple regnancies involve a igher chance of remature birth.

lentical twins from multiple pregnancy

Fraternal twins

Twins resulting from two fertilized eggs that develop simultaneously

Sometimes a woman may produce two egg cells, or ova, which are fertilized at the same time by different sperm ▪. If both eggs develop, the woman will give birth to fraternal twins. They are also known as **dizygotic twins**, because they develop from two different zygotes ▪. The twins each have a different set of genes ▪, and can be different sexes. They are only as similar as any other brother or sister and may look quite different.

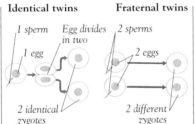

Two separate placentas

One shared placenta

Identical twins   **Fraternal twins**

1 sperm *Egg divides in two* *2 sperms*

1 egg   *2 eggs*

2 identical zygotes   *2 different zygotes*

Identical and fraternal twins
Identical twins come from one divided egg. Fraternal twins come from two eggs.

Identical twins

Twins resulting from one fertilized egg that divides in two

Occasionally, a fertilized egg cell splits into two cells that develop separately. If each new cell survives, they develop into identical twins, or **monozygotic twins**. They have exactly the same genes and are always the same sex. They usually look so similar that they are difficult to tell apart.

Infertility

An inability to conceive

Sexual reproduction is a complicated process, with many separate steps. If any of the steps fails to work, the result may be infertility. Infertility can result from problems in either partner. Men may be infertile because they cannot produce enough sperm, while women may be infertile because they do not ovulate ▪ normally. Sometimes, infertility can be treated by the technique of **in vitro fertilization** (in vitro means "in glass"). This is where eggs and sperm are brought together in a test tube, and fertilized eggs are then placed in the woman's uterus.

Birth control

The deliberate control of pregnancy and birth

Birth control, or **family planning**, is a variety of measures that couples use to limit the number of children they have. The simplest method of birth control involves not having sexual intercourse ▪, or avoiding it when a woman is at her most fertile. Many other methods involve using contraceptives.

Contraceptive

Something designed to prevent conception

A contraceptive allows a couple to have sexual intercourse, but stops a pregnancy from resulting. The simplest kinds are **barrier contraceptives**, such as **condoms**, which prevent eggs and sperm coming together. **Hormonal contraceptives**, the most common of which is the **pill**, contain small amounts of female sex hormones ▪. They usually work by preventing ovulation. Some contraceptives are quite reliable, but none has a 100 percent success rate.

Birth

Birth is a dramatic event for mother and baby alike. After nine months in the warmth and security of its mother's body, a baby suddenly emerges into a very different world.

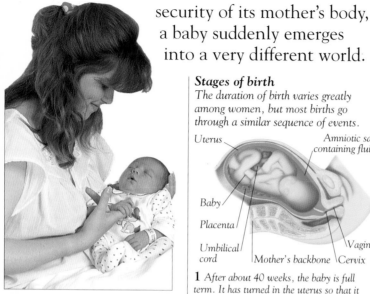

Mother with newborn baby
This woman has recently given birth to her new baby.

Birth

The process that carries a child out of its mother's uterus and into the world outside

Birth is also known as **parturition**. It normally occurs after 38–42 weeks of pregnancy ▣, and is triggered by changes in the levels of hormones ▣ in the mother's blood. During birth, the uterus contracts, and the baby is squeezed head-first through the mother's birth canal, or vagina ▣. There is very little room to spare, and the baby's head is temporarily squashed out of shape as it passes through the mother's pelvis ▣. Once the baby has emerged, it soon takes its first breath, and blood flows through its lungs for the first time. A few minutes after the baby has been born, the afterbirth, or placenta ▣, is also expelled from the uterus ▣.

Stages of birth
The duration of birth varies greatly among women, but most births go through a similar sequence of events.

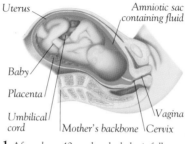

Uterus — Amniotic sac containing fluid
Baby
Placenta
Umbilical cord — Vagina
Mother's backbone — Cervix

1 *After about 40 weeks, the baby is full term. It has turned in the uterus so that it is facing head down, ready to be born.*

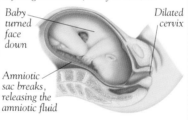

Baby turned face down — Dilated cervix
Amniotic sac breaks, releasing the amniotic fluid

2 *During the first stage of labor, the cervix widens to a diameter of about 4 inches (10 cm).*

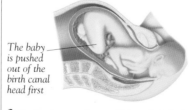

The baby is pushed out of the birth canal head first

3 *During the second stage of labor, the baby is born. Contractions push the baby out of the uterus and through the birth canal, or vagina.*

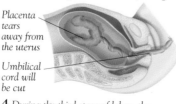

Placenta tears away from the uterus
Umbilical cord will be cut

4 *During the third stage of labor, the placenta is expelled from the body.*

Labor

A period of muscle contractions before and during birth

Labor usually begins about 12 hours before birth occurs, although it may begin later if the mother has already had other children. During the **first stage of labor**, the uterus starts to contract rhythmically, and the cervix ▣ slowly opens. During the **second stage of labor**, the top of the baby's head becomes visible. After this, powerful **contractions** push the baby out of the uterus and out through the vagina. In the **third stage of labor**, the placenta is expelled. Labor is triggered by the hormone oxytocin ▣, which is released by the mother's pituitary gland ▣.

Breech birth

A birth in which the baby is born head upward

In the early stages of pregnancy, a developing fetus ▣ is normally positioned head upward. By about week 32, it has usually turned upside down to allow its widest part – the head – to be born first. In a breech birth, the baby is still head upward, and is born legs and bottom first. A breech birth carries a greater risk of injury to the baby, and is often avoided by carrying out a Cesarean section.

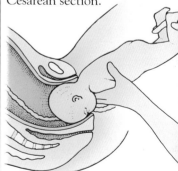

Breech birth
During a breech birth, a baby is born legs and bottom first, rather than head first.

Cesarean section

The delivery of a baby through an incision in the mother's abdomen

A Cesarean section is a surgical operation in which a baby is removed directly from its mother's uterus. It gets its name from the Latin word meaning "to cut." It is carried out if a baby cannot fit through its mother's pelvis, or if another complication arises during pregnancy or labor. During the operation, a short horizontal slit is cut through the mother's lower abdomen. The baby and placenta are taken out and the slit is sewn up. In some parts of the world, large numbers of babies are born in this way.

Premature baby
This tiny baby is being cared for in the protective environment of an incubator.

Premature birth

A birth that occurs after less than 37 weeks of pregnancy

In some parts of the world, many babies are born before they have fully developed. Premature babies often have difficulty in breathing due to a lack of surfactant ▪. They also lack the strong sucking reflex that enables full-term babies to feed soon after they are born. At one time, most premature babies died. Today, they are cared for in incubators, which are transparent containers that hold moist air at body temperature. Incubators can even enable babies born more than 12 weeks early to survive.

Birth defect

A defect noticeable at birth or during infancy

There is no such thing as a "perfect" baby, because everyone is born with blemishes of some kind. Some birth defects are serious enough to need medical attention. Birth defects may be caused by infections that the mother has suffered, by medicines taken during pregnancy, or as a result of genetic disorders ▪, which stop the embryo ▪ or fetus from developing normally. A **birthmark** is a harmless region of discolored skin that forms when the baby is developing inside its mother. Some birthmarks are permanent; others disappear during childhood.

Miscarriage

The loss of a baby before it can survive outside its mother's body

About 20 percent of pregnancies result in the embryo or fetus dying during development. Most miscarriages occur during the first 12 weeks of pregnancy, and they have many causes. The embryo or fetus may die because it has a built-in genetic defect. In other cases, miscarriages are caused by an illness that the mother has suffered. A miscarriage does not usually affect the woman's ability to have children in the future, but it is a traumatic experience.

Milk

A liquid food produced by a mother

Human milk, or **breast milk**, is a complete food that nourishes a baby for the first few months of its life. It is rich in a sugar called lactose ▪, and it also contains fats, proteins, and small amounts of antibodies ▪.

See also

Antibody 98 • Cervix 129
Embryo 137 • Fetus 137
Genetic disorder 135 • Hormone 78
Lactose 22 • Lymphocyte 83
Oxytocin 79 • Pelvis 41
Pituitary gland 79 • Placenta 138
Pregnancy 138 • Surfactant 113
Uterus 129 • Vagina 129

Colostrum

A watery fluid produced before milk

During late pregnancy, and just after birth, a mother's breasts produce a yellow fluid called colostrum. Colostrum is rich in antibodies and lymphocytes ▪, which help a baby to defend itself against infection. It is replaced by milk about a week after birth.

Lactation

The process of producing milk

Milk is produced by a mother's **breasts**, or **mammary glands**. Each breast contains 15–20 clusters of milk-secreting glands called **lobes**, and has a raised center called a **nipple**. The glands are connected to the nipple by tubes called **milk ducts**. During breast-feeding, a baby triggers its mother's breasts to release milk through the nipple. This is the **milk let-down reflex**, which takes about a minute to work and involves the hormone oxytocin.

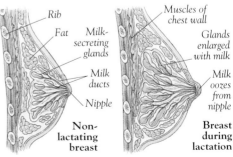

Rib
Fat
Milk-secreting glands
Milk ducts
Nipple
Non-lactating breast

Muscles of chest wall
Glands enlarged with milk
Milk oozes from nipple
Breast during lactation

Lactation
The milk-secreting glands in breasts get larger and produce milk during lactation.

Growth & aging

Throughout life, the human body follows a predictable pattern of changes. Infants become children, children become adolescents, and adolescents become adults. From middle age onward, the body begins to show signs of aging.

Growth

An increase in size

Growth occurs mainly through cell division ■. It is controlled by a number of factors, including hormones ■ and diet ■. Instead of growing steadily throughout life, the body grows at different rates at different times. The fastest burst of growth occurs before birth ■. Growth is still rapid during infancy, but it slows down in childhood. A second spurt of growth occurs during puberty. In adulthood, growth ceases completely, although cells continue to divide.

1 An infant has a large head, and short arms and legs. It relies completely on its parents.

Infant

A baby under 12 months old

The first stage of a baby's life outside the mother's body is called **infancy**. An infant is helpless and relies on its parents for food and protection.

2 A two-year-old child has much longer arms and legs than an infant. Its legs are now strong enough for walking. It is less helpless than an infant.

Development

An increase in complexity

From a single cell, called a zygote ■, many different kinds of cell develop to form a human body. Development involves the differentiation ■ of cells, which tailors them to different tasks. It also involves **differential growth**, which makes some parts of the body grow faster than others. This kind of growth makes the body change shape. Development and growth usually go hand in hand, but sometimes one can occur without the other. For example, development initially happens without growth once a zygote has been formed.

Adolescence

The period that separates childhood and adulthood

Adolescence occurs during the teenage years. It is a period of physical and psychological changes that prepare the body for adulthood. During this period, young people learn how to live independently of his or her parents, and how to deal with the beginning of sexual maturity. Adolescence can be a difficult time. Many adolescents worry about their physical appearance, and their growing need for independence can often bring about disagreements with their parents.

Puberty

The period of physical change during adolescence

Puberty is a period of rapid growth and development that is triggered by sex hormones ■. It lasts for about three years. Puberty usually begins at the age of about 11 years in girls, and about 2 years later in boys, although the exact age varies from one person to another. During puberty, the body develops secondary sexual characteristics ■ that prepare it for sexual reproduction ■.

3 By about the age of 13 years, the body is growing and changing as a result of puberty. Body hair begins to grow and in girls, the breasts develop, and the hips widen.

4 By about the age of 18 years, most people are fully grown. Only structures such as the wisdom teeth are still growing.

Growing up

Growth and development follow similar patterns in all children. A final spurt of growth typically occurs during puberty, which is between the ages of about 11 and 13 years in girls and 13 and 16 years in boys.

Adult

A person who has completed his or her growth and development

By about 18 years of age, the body stops growing. This marks the beginning of adulthood. In early adulthood, the body may still be developing, but only in minor ways. These late changes include the ossification ■ of parts of the breastbone, or sternum ■, and the eruption of wisdom teeth ■.

ging

process of growing older

ing is also known as
escence. It is a process that
ins around middle age, and
cts all the body's cells, making
m less efficient at carrying out
ryday tasks. As a result, the
ly as a whole begins to
nge. The skin ■ becomes less
stic and muscles lose some of
ir strength, while the bones ■
y become brittle and joints
s mobile. Scientists do not yet
w just why aging occurs.
nay be because cells
wly lose the ability
divide, or because
y build up chemical
-products, or
tabolites ■, which
come toxic. Aging affects
ple in different ways,
l is almost certainly
uenced by genes ■.
wever, regular
ercise and a balanced
t can help to
nimize its effects.

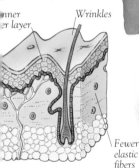

nner
er layer

Wrinkles

*Older skin is thinner, and has
er elastic fibers in its deeper
ers. As a result, the skin appears
er, and becomes more wrinkled.*

*Fewer
elastic
fibers*

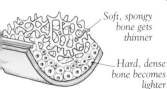

*Soft, spongy
bone gets
thinner*

*Hard, dense
bone becomes
lighter*

*In older people, bone has less
agen, which gives it elasticity,
l less calcium, which gives it
ength. As a result, the bone
omes thinner and more brittle.*

Death

The permanent end of the body's
life processes

When someone dies, one or
more of their body systems
breaks down. As a result, the
rest of the body can no longer
work, and its chemical processes
come to a permanent halt. This
change does not happen in an
instant, and it can sometimes be
reversed if it has not progressed
too far. During **resuscitation**, a
person's breathing and
heartbeat are artificially
restarted, bringing the rest
of the body back to life.

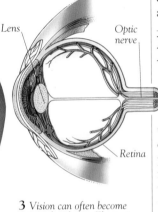

Lens

*Optic
nerve*

Retina

3 *Vision can often become
impaired with age. Elastic tissue
in the lens stiffens, so that it is
unable to change shape. This
affects the ability to focus.*

Clot

*Lining of
artery gets
thinner*

Plaque developing

4 *Layers of fatty material, called
plaque, can build up inside an
artery, making it narrower.
Blood clots may form, causing
strokes and heart attacks.*

Signs of aging

*As the body ages, a number
of physical changes occur.
Some are inevitable, but
others can be avoided by a
healthy life-style.*

Life expectancy

The number of years that a newly
born baby can expect to live

On average, humans are able to
live for between 75 and 85 years
of age. This is their natural **life
span**. However, the age people
can actually expect to reach is
usually lower than this because
of the effects of disease, and of
other hazards in everyday life.
The average life expectancy in
the developed (industrialized)
world is between 70 and 75 years
of age. In the developing
(nonindustrialized) world, it is
often much less. This is mainly
due to a lack of adequate food
and health care.

Mortality rate

The number of deaths per year in a
group of people

The mortality rate is also
called the **death rate**. It is
usually estimated as the number
of deaths each year per 1,000
people. The **birth rate** is the
number of children born each
year per 1,000 people. If birth
rate and death rate are the same,
the overall number of people, or
population, stays the same.
During the past century,
birth rates in most parts
of the world have
exceeded death
rates, so the human
population has
grown rapidly.

See also

Infectious diseases

This table lists important diseases that are caused by infectious organisms such as bacteria and viruses. Many are highly contagious, which means that they can easily be "caught" from an infected person. Others, such as malaria, cannot be spread directly from one person to another.

AIDS (*see page 99*)

Amoebic dysentery (Amoebiasis)
Causative agent: Protozoan (*Entamoeba histolytica*)
Transmission: Spreads through food, especially when uncooked, and water contaminated by feces.
Effects: Multiplies in the large intestine, and can cause diarrhea and bleeding of intestinal wall.

Chickenpox
Causative agent: Virus (Herpesvirus)
Transmission: Spreads through airborne droplets.
Effects: Causes a widespread skin rash that develops into blisters. This common childhood disease is rarer and more serious in adults.

Cholera
Causative agent: Bacterium (*Vibrio cholerae*)
Transmission: Spreads through contaminated water or food.
Effects: Causes vomiting and violent diarrhea. Without treatment, people with this disease can die from dehydration.

Common cold
Causative agent: Virus (Rhinoviruses and others)
Transmission: Spreads through airborne droplets.
Effects: Causes inflammation of the nasal sinuses and the mucous membranes lining the throat. It is actually a number of diseases, caused by over 200 viruses.

Diphtheria
Causative agent: Bacterium (*Corynebacterium diphtheriae*)
Transmission: Spreads through airborne droplets, or by touch.
Effects: Infects the throat and trachea, and in severe cases can cause paralysis. In many countries, childhood immunization has made this disease less common.

Dysentery *see* **Amoebic dysentery**

German measles (Rubella)
Causative agent: Virus (Togavirus)
Transmission: Spreads through airborne droplets.
Effects: Causes a red rash. Usually a mild disease, but it can cause birth defects when it infects pregnant women. Can be prevented by immunization.

Giardiasis
Causative agent: Protozoan (*Giardia lamblia*)
Transmission: Spreads through infected water or food, and through sexual contact.
Effects: Multiplies in the small intestine, causing intestinal cramps and severe diarrhea. It is most common in the tropics.

Glandular fever (Mononucleosis, infectious)
Causative agent: Virus (Herpesvirus)
Transmission: Uncertain, although possibly via saliva.
Effects: Swollen lymph glands, tonsillitis, headache, lethargy; may last for several weeks.

Gonorrhea
Causative agent: Bacterium (*Neisseria gonorrhoeae*)
Transmission: Spreads during sexual intercourse.
Effects: Causes inflammation of the sexual organs. Sometimes spreads to other body parts.

Hepatitis
Causative agent: Virus (various types)
Transmission: Hepatitis A spreads by contaminated food and water, or physical contact. Hepatitis B spreads by infected blood, and sexual intercourse.
Effects: Involves inflammation of the liver. This can cause jaundice (yellowing of the skin and eyes), and sometimes severe liver disease.

Influenza (Flu)
Causative agent: Virus (various types)
Transmission: Spreads through airborne droplets.
Effects: Causes headache, fever and aching joints. Many strains of influenza exist, making it hard for the body to develop immunity.

Leprosy
Causative agent: Bacterium (*Mycobacterium leprae*)
Transmission: Spreads through prolonged close contact.
Effects: Infects the peripheral nervous system, causing loss of sensation. This can lead to damage through accidental injuries. It is difficult to cure and is widespread in parts of Africa and Asia.

Malaria
Causative agent: Protozoan (*Plasmodium* species)
Transmission: By mosquito bite.
Effects: Attacks red blood cells and causes repeated bouts of chill and fever; may cause liver and kidney failure. This widespread and dangerous disease can be treated with antimalarial drugs.

Measles
Causative agent: Virus (Paramyxovirus)
Transmission: Spreads rapidly through airborne droplets.
Effects: Causes skin rashes and fever. This widespread disease mainly affects children, but can be prevented by immunization.

Meningitis
Causative agent: Virus or bacterium (various kinds)
Transmission: Causative agents are often present in healthy people, but only cause disease when they are able to reach the central nervous system.
Effects: Causes inflammation of the meninges (membranes around the brain and spinal cord), leading to headache and fever. Some forms of bacterial meningitis may be very dangerous and cause brain damage or death.

Mumps
Causative agent: Virus (Paramyxovirus)
Transmission: Spreads through airborne droplets.
Effects: Causes swelling of the salivary glands. This disease affects children and adults, but can be prevented by immunization.

Pneumonia
Causative agent: Virus or bacterium (various types)
Transmission: Usually spreads through airborne droplets.
Effects: Causes inflammation of the lungs, sometimes making breathing difficult. Pneumonia often accompanies other diseases.

Poliomyelitis (Polio)
Causative agent: Virus (Poliovirus)
Transmission: Spreads mainly through contaminated water.
Effects: Range from headache and fever to partial paralysis. This potentially dangerous disease can be prevented by immunization.

Rabies
Causative agent: Virus (Rhabdovirus)
Transmission: Through the bite of an infected animal, such as a dog, fox, or raccoon.
Effects: Fever, delirium, muscle cramps in throat. Immunization given within two days of an animal bite usually prevents rabies.

Shingles (Herpes zoster)
Causative agent: Virus (Herpesvirus)
Transmission: Spreads through airborne droplets.
Effects: Caused by the same virus that produces chickenpox and usually affects people over 50. The virus infects nerves. Years after having chickenpox, it produces a rash on one side of the body.

Smallpox
Causative agent: Virus (Poxvirus)
Transmission: Spreads through airborne droplets.
Effects: Causes large blisters, or lesions, on the skin, and is often fatal. Once widespread, smallpox has been eradicated by a worldwide vaccination program.

Syphilis
Causative agent: Bacterium (Treponema pallidum)
Transmission: Spreads through sexual contact.
Effects: Initially causes sores and rashes. If left untreated, it can lead to organ damage and insanity. It can last for years, and may be passed on to children before birth.

Tetanus
Causative agent: Bacterium (Clostridium tetani)
Transmission: Spreads mainly by contact with soil and manure contaminated with bacteria. The bacteria may then enter deep cuts.
Effects: Infects central nervous system, making muscles rigid and later stopping breathing. It can be prevented by immunization.

Tuberculosis
Causative agent: Bacterium (Mycobacterium tuberculosis)
Transmission: Spreads through airborne droplets and cow's milk.
Effects: Infects the lungs and sometimes other organs, causing small lumps called tubercles. Can be treated, but kills many people each year in developing countries.

Typhoid fever
Causative agent: Bacterium (Salmonella typhi)
Transmission: Spreads through water contaminated by feces.
Effects: Causes headaches, fever, and sometimes severe disorders of the digestive system. It is common in countries with poor sanitation.

Typhus
Causative agent: Bacterialike organisms (Rickettsiae)
Transmission: Through the bites of lice, ticks, and other animals.
Effects: This serious disease can produce fever, blood poisoning, pneumonia, and heart failure. Normally uncommon, it can flare up after natural disasters, when people are crowded together.

Whooping cough (pertussis)
Causative agent: Bacterium (Bordetella pertussis)
Transmission: Spreads through airborne droplets.
Effects: Causes inflammation of the trachea and airways of lungs, and severe coughing. This disease mainly affects children, and can be prevented by immunization.

Yellow fever
Causative agent: Virus (Togavirus)
Transmission: By mosquito bites.
Effects: Infects lymphatic system and many internal organs; causes liver damage and jaundice – yellowing of the skin and eyes. It is widespread in the American and African tropics, but can be prevented by immunization.

Noninfectious diseases & disorders

This table lists important diseases and disorders that are not caused by infectious organisms. Some are the result of inheriting defective genes. Others are triggered by environmental factors such as allergens in the air.

Albinism
Cause: Inherited disorder involving defective gene(s).
Effects: Inability to produce the pigment melanin. People with albinism have pale skin and white hair, and are easily sunburned.

Allergy (*see page 100*)

Alzheimer's disease
Cause: Unknown. Possible causes include poisoning by aluminum, and the action of inherited genes.
Effects: Loss of memory, mental confusion, lack of coordination. This disease usually strikes people who are more than 60 years old, and at present cannot be cured.

Anemia (*see page 82*)

Anorexia nervosa (*see page 107*)

Arthritis (*see page 47*)

Asbestosis
Cause: Prolonged exposure to fibers of asbestos, a mineral that was formerly used as a building material.
Effects: Scarring of the lungs, causing difficulty in breathing; can eventually lead to severe disability and death. Asbestosis is one of several lung diseases that are caused by dust or fibers in the air.

Asthma
Cause: Usually an allergic reaction to substances in the air, or a response to sudden changes in air temperature.
Effects: Narrowing of bronchioles in the lungs, causing difficulty in breathing. Common in children.

Atherosclerosis (*see page 89*)

Beriberi (*see page 109*)

Cancer (*see page 31*)

Celiac disease
Cause: An allergic reaction to gluten, a protein found in wheat and some cereals.
Effects: Damage to the lining of the small intestine, leading to weakness and fatigue. Can be treated with a gluten-free diet.

Color blindness (*see page 71*)

Deafness (*see page 73*)

Diabetes
Cause: An inability to produce sufficient quantities of the hormone insulin. This can be triggered by an infection of the pancreas or by genetic factors.
Effects: Abnormally high levels of glucose in the blood, leading to weakness and fatigue, reduced effectiveness of immune system, and sometimes coma. Treated by regular injections of insulin or other drugs and controlled sugar intake.

Down syndrome
Cause: Chromosome abnormality; people with Down syndrome have three copies of chromosome 21, instead of the normal two.
Effects: Characteristic facial features, including a fold of skin on both sides of the nose; varying degrees of mental handicap.

Dwarfism (*see page 79*)

Eczema
Cause: Several, including allergies, irritation produced by contact with chemicals, and the action of inherited genes.
Effects: Skin inflammation, often resulting in sore patches and dry, scaly areas.

Emphysema
Cause: Smoking; in rare cases, the action of inherited genes.
Effects: Damage to the walls of the alveoli in the lungs. This leads to breathlessness, and to complications caused by insufficient oxygen in the blood.

Epilepsy
Cause: Disorders in the normal activity of the brain; these may be produced by infections, by drugs, or by the action of inherited genes.
Effects: These vary from short-lived loss of consciousness and seizures to temporary paralysis of the whole body.

Farsightedness (*see page 70*)

Gigantism (*see page 79*)

Hemophilia (*see page 84*)

Heart attack (*see page 87*)

Hyperthyroidism
Cause: Excessive activity of the thyroid gland.
Effects: Increased metabolic rate and overactivity.

Kwashiorkor
Cause: Malnutrition.
Effects: Slowed growth, enlargement of the liver, flaking skin, weakness. Kwashiorkor mainly affects young children, usually when they start to eat solid food. It is common in parts of the world where there are severe food shortages.

Leukemia
Cause: Uncertain; involves rapid multiplication of white blood cells in the bone marrow.
Effects: Leukemia is a form of cancer. It can reduce the body's ability to fight infections, and causes tiredness and bruising of the skin. Often affects children.

Migraine
Cause: Uncertain; may be inherited, but is often triggered by particular types of food, and by environmental factors, such as bright lights and loud noise.
Effects: Severe headache, often with vomiting and sensitivity to light.

Motor neuron disease
Cause: Unknown, but the most common forms of the disease seem to be inherited.
Effects: Triggers the breakdown of motor neurons in the brain and spinal cord, which leads to gradual loss of movement. Mainly affects people over 50, and is more likely to occur in men than women.

Multiple sclerosis (MS)
Cause: Unknown, but probably involves the immune system disorder in which the body attacks its own nervous tissue.
Effects: Gradual destruction of areas of myelin in the brain and spinal cord. This produces numbness, weakness, and loss of muscle control. MS usually affects young adults and can persist throughout life.

Muscular dystrophy
Cause: Action of inherited genes.
Effects: Gradual breakdown of muscle fibers, causing weakness and sometimes an inability to move about. There are several forms of muscular dystrophy; those triggered by sex-linked genes occur only in males.

Nearsightedness (*see page 70*)

Night blindness (*see page 71*)

Obesity (*see page 107*)

Osteoporosis (*see page 35*)

Parkinson's disease
Cause: Unknown.
Effects: In this disease, some of the brain's cells degenerate. This affects muscle control, causing tremor (shaky movement), stiffness, and weakness. Mainly affects people over 60 years of age.

Phenylketonuria (PKU)
Cause: Inherited disorder involving a single defective gene.
Effects: Prevents the body from breaking down the amino acid phenylalanine in the normal way. If this amino acid is allowed to build up, it can produce epilepsy and mental disorders. PKU is one of several hundred metabolic disorders that block particular chemical reactions in the body.

Psoriasis
Cause: Unknown, but in many cases it is an inherited disorder. Usually occurs in sudden attacks caused by illness or stress.
Effects: Inflammation of the skin, with rapid division of the cells just beneath the surface. This may produce red spots or larger scaly patches. In severe cases, is accompanied by arthritis.

Rheumatism (*see page 47*)

Rheumatoid arthritis (*see page 47*)

Scurvy (*see page 109*)

Schizophrenia
Cause: Uncertain. Can be inherited; may also be caused by brain damage or the action of some drugs.
Effects: Schizophrenia means "split mind." It affects the central nervous system, particularly in young adults. People who suffer from schizophrenia often have severe emotional disturbance and may behave illogically.

Sciatica
Cause: Pressure on the sciatic nerve, usually from a prolapsed intervertebral disk.
Effects: Pain extending down the leg, muscle weakness.

Sickle-cell anemia
Cause: An inherited gene that produces an abnormal type of hemoglobin.
Effects: Red blood cells develop a sickle (crescent) shape, which reduces their ability to carry oxygen around the body. This disease can produce fatigue and may damage internal organs. It mainly affects black people.

Spina bifida
Cause: Unknown, but occurs most often in children with particularly young or old mothers.
Effects: Spina bifida means "spine divided in two." In this disorder, a fetus's spinal cord fails to develop normally, and is not completely enclosed within the backbone. In serious cases, it produces a permanent handicap.

Stroke (*see page 84*)

Tinnitus
Cause: Damage to the acoustic nerve; often a symptom of other disorders.
Effects: Persistent whistling or ringing sound in the ear that is not produced by real sound waves.

Pioneers of human biology & medicine

Addison, Thomas
British doctor (1793–1860)
Investigated the adrenal glands, and helped to found the science of endocrinology.

Alcmaeon
Greek doctor & philosopher (born c.535 BC)
Discovered the optic nerve, and realized that the brain – rather than the heart – is the organ involved in feeling, sensations, and thinking.

Alhazen
Arab physicist (c.965–1038)
Studied the physics of light and vision, or optics, and realized that eyes absorb rather than give out light rays, as was previously thought.

Anderson, Elizabeth Garrett
British doctor (1836–1917)
Became the first woman to qualify as a doctor in Great Britain, breaking the tradition that only men could practice medicine.

An-Nafis, Ibn
Arab doctor & anatomist (died 1288)
First person to show that blood flows through the lungs, forming a circulatory system.

Avicenna, Ibn Sina
Arab philosopher & doctor (980–1037)
Wrote *The Canon of Medicine*, a multivolume medical textbook that remained in use in Europe for over 500 years.

Baer, Karl von
Estonian embryologist (1792–1876)
Founder of modern embryology. Discovered that a Graafian follicle contains an ovum, and investigated the development of animal form.

Banting, Sir Frederick
Canadian physiologist (1891–1941)
Devised a method of obtaining insulin from the pancreas, providing a way to control the effects of diabetes.

Bartholin, Caspar
Danish anatomist (1585–1629)
In 1611, published *Anatomicae Institutiones Corporis Humani*, or "Anatomical Principles of the Human Body," a major anatomical textbook. His son Thomas Bartholin (1616–1680), who was also a distinguished anatomist, realized that lymph vessels form a separate body system.

Beaumont, William
American surgeon (1785–1853)
Researched the mechanisms behind digestion, working with the help of a man who had been injured after a shooting accident. The injury to the man's stomach allowed Beaumont to collect samples of its contents.

Bell, Sir Charles
British surgeon and neurologist (1774–1842)
Showed that nerves contain many separate fibers (called neurons), and discovered that each fiber carries either sensory or motor signals, but not both. Also deduced that most muscles must be supplied with both sensory and motor fibers.

Benenden, Edouard van
Belgian cytologist (1846–1910)
Discovered that every species, including the human species, has a characteristic number of chromosomes in its cells.

Bernard, Claude
(*See page 77*)

Bichat, Marie François
French pathologist (1771–1802)
Showed that organs are made of different groups of cells, which he named "tissues." Helped to lay the foundations of histology.

Bowman, Sir William
British doctor (1816–92)
Made detailed studies of the fine structure of tissues, particularly in the muscles and the kidneys. Bowman's capsule is named after him.

Broca, Paul
French surgeon (1824–80)
First person to show that particular regions of the brain control specific functions of the body. He realized this after discovering that a man who had been unable to talk had suffered damage to a small part of his brain. Now known as Broca's area, this region controls speech.

Buchner, Eduard
German chemist (1860–1917)
Showed that fermentation could occur outside living cells. This in turn led to the understanding of enzymes.

Calmette, Albert
French bacteriologist (1863–1933)
With his associate Camille Guérin, developed a vaccine (BCG) that confers immunity to tuberculosis.

Chain, Sir Ernst Boris
German-British biochemist (1906–79)
Helped to isolate penicillin so that it could be used as a drug.

Chargaff, Erwin
American biochemist (born 1905)
Showed how the four different base pair up in a DNA molecule.

Chen Chuan
Chinese doctor (died 643)
The first person known to have recorded the symptoms of diabetes. During his time, Chinese medicine was more advanced than medicine in the Western world.

Colombo, Matteo
Italian anatomist (1516–59)
Demonstrated that blood flows from the heart to the lungs, and then back again.

Crick, Francis
British biochemist (born 1916)
With the American biochemist James Watson, built a model that showed the double helix shape of the DNA molecule for the first time. This was one of the most important discoveries in modern science.

Dioscorides, Pedanius
Greek doctor (born c.20)
Wrote *De Materia Medica*, a compendium that dealt with the medical uses of hundreds of different plants.

oisy, Edward
merican biochemist (1893–1986)
olated vitamin K, which plays a
art in the clotting of blood.

uve, Christian de
elgian biochemist (born 1917)
iscovered lysosomes, the organelles
at cells use to digest substances or
ometimes themselves.

hrlich, Paul
See page 93)

ijkman, Christiaan
utch doctor (1858–1930)
iscovered that the deficiency
isease beriberi can be cured by
change in diet.

inthoven, Willem
utch physiologist (1860–1927)
vented the electrocardiograph
ECG), a device for monitoring
e activity of the heart.

mpedocles
reek doctor and philosopher
ied c.430 BC)
ne of the first people to realize that
e heart is at the center of a system
f blood vessels. At this time, the
ea of circulation was still unknown.

ustachio, Bartolommeo
alian anatomist (1524–74)
tudied the structure of several
rgans and systems, including the
idney, the ear, and the sympathetic
ervous system. First person to
escribe Eustachian tubes.

abricius, Hieronymus
alian anatomist (1537–1619)
n 1603, published *De Venarum
Ostiolis*, or "On the Valves of Veins,"
he first clear description of veins.
Also laid the foundations of the
cience of embryology.

allopio, Gabriello (Fallopius)
alian anatomist (1523–62)
iscovered the tubes that
onnect the ovaries with the
terus (Fallopian tubes).

ick, Adolf
See page 113)

leming, Sir Alexander
ritish microbiologist (1881–1955)
While working with bacteria,
oticed that a particular mold
eemed to kill them. This led to
he discovery of penicillin, the
rst antibiotic.

Flemming, Walther
German cell biologist
(1843–1905)
Was the first person to observe
and describe the separate stages
of mitosis during cell division.

Florey, Sir Howard
Australian pathologist (1898–1968)
Helped to isolate the antibiotic
penicillin, so that it could be
used as a drug.

Fracastoro, Girolamo
Italian doctor (1478–1553)
In 1546, published *De Contagione
et Contagiosis Morbis*, or "On
Contagion and Contagious
Diseases," an early attempt to
explain how infectious diseases
are spread. He suggested that
these diseases are spread by
seedlike particles – an idea not
far from the truth.

Franklin, Rosalind
British biochemist (1920–58)
Used X-rays to investigate the
shape of DNA molecules. Her
research helped reveal its double
helix structure.

Galen, Claudius (Galenus)
Greek anatomist (129–199)
Investigated the structure and
function of the human body. His
ideas – some of them mistaken –
stayed in use throughout Europe
for many centuries.

Galvani, Luigi
Italian anatomist (1737–98)
Accidentally discovered that
electricity can make muscles
contract.

Golgi, Camillo
(*See page 59*)

Graaf, Regnier de
(*See page 131*)

Hales, Stephen
British physiologist & chemist
(1677–1761)
Became the first person to measure
blood pressure, using a horse; also
measured blood flow and showed
that capillaries can change shape by
constricting (becoming narrower).

Haller, Albrecht von
Swiss physiologist (1708–77)
Helped to found the science of
neurology, showing that nerves carry
sensory information to the brain.

Harvey, William
British doctor (1578–1657)
Published *De Motus Cordis*, or
"On the Movement of the Heart,"
the first full account of how blood
circulates around the body. He
correctly concluded that blood must
pass from arteries to veins, although
he did not know of the capillaries
that make it possible.

Havers, Clopton
(*See page 35*)

Helmholtz, Hermann
German physicist & physiologist
(1821–94)
As well as making many discoveries
in physics, Helmholtz invented the
ophthalmoscope, an instrument for
examining the retina in the eye.

Henle, Friedrich
German anatomist (1809–85)
Studied contagious diseases and also
helped to establish the science of
histology. The loop of Henle, in
nephrons of the kidney, is named
after him.

Hippocrates
Greek doctor (c.460–377 BC)
Founder of the science of medicine,
which relies on informed diagnosis,
rather than myth and magic.

His, Wilhelm
Swiss histologist (1831–1904)
Inventor of the microtome – a
device that cuts very thin slices of
tissue, which can be examined easily
under a microscope.

Hodgkin, Dorothy
British biochemist (born 1910)
Developed a technique for
investigating the structure of
molecules by firing X-rays at pure
crystals; she used this to work out
the structure of penicillin and
vitamin B_{12}.

Hooke, Robert
English physicist & microscopist
(1635–1703)
First person to identify cells; in
1665, published *Micrographia*, an
illustrated survey of objects seen
under the microscope.

Hopkins, Frederick Gowland
British biochemist (1861–1947)
Carried out wide-ranging
experiments concerning the effects
of vitamins, and showed that they
are an essential part of the diet.

Continued over page ➤

Hunter, John
British surgeon & anatomist (1728–93)
Made advances in the surgical treatment of wounds and published one of the earliest works on sexually transmitted diseases. Also amassed a large collection of anatomical specimens, including the skeleton of a giant man, who was 7 feet 7 inches (2.3 m) tall.

Jenner, Edward
(*See page 101*)

Khorana, Har Gobind
Indian-American biochemist (born 1922)
Helped to crack the genetic code by identifying which amino acid is specified by all the possible codons (combinations of three nucleic acid bases).

Kitasato, Shibasaburo
Japanese bacteriologist (1852–1931)
One of the discoverers of the plague bacterium Pasteurella pestis in 1894.

Koch, Robert
German bacteriologist (1843–1910)
While studying anthrax, a disease found in cattle and people, became the first person to prove that bacteria can cause disease. He also discovered the bacterium that produces tuberculosis.

Krebs, Hans
German biochemist (1900–81)
Discovered the sequence of chemical reactions (Krebs cycle) in aerobic respiration that breaks down glucose to release energy.

Kühne, Wilhelm
German physiologist (1837–1900)
First person to use the term enzyme; discovered that visual pigments undergo a chemical change when exposed to light.

Laennec, René Théophile
French doctor & surgeon (1781–1826)
Studied diseases of the chest and devised the first stethoscope, using a hollow tube and a wooden rod.

Landsteiner, Karl
(*See page 85*)

Langerhans, Paul
German doctor (1847–88)
Discovered the groups of cells in the pancreas, which are now known as islets of Langerhans.

Lavoisier, Antoine
(*See page 105*)

Leeuwenhoek, Antony van
(*See page 129*)

Levi-Montalcini, Rita
Italian neurophysiologist (born 1909)
Investigated the way nerves develop, and discovered that many more neurons are produced than are needed. The redundant cells die during development.

Lind, James
British doctor (1716–94)
Found that citrus fruit prevents the deficiency disease scurvy.

Lister, Joseph
British doctor & surgeon (1827–1912)
Introduced the use of antiseptics in surgery. As a result, deaths following amputations were reduced by half.

Lower, Richard
British physiologist (1631–91)
Carried out the first successful blood transfusion using animals, and investigated the function of the heart and lungs. Discovered that blood from veins turns bright red when it comes into contact with air.

Ludwig, Karl
German physiologist (1816–95)
Helped to found the science of physiology by using known physical principles to explain body processes.

Macewen, Sir William
British surgeon (1848–1924)
Demonstrated the value of aseptic techniques in preventing infections after surgery.

Malpighi, Marcello
Italian biologist (1628–94)
Became the first person to see capillaries, and investigated other tissues, including nerves and the skin. The Malpighian layer in the skin is named after him.

Mead, Margaret
American anthropologist (1901–78)
Carried out detailed studies of human behavior in different societies.

Mechnikov, Ilya
Russian-French bacteriologist (1845–1916)
Discovered phagocytosis in animal cells, and found that it is also carried out by white cells in human blood.

Medawar, Peter
British immunologist (1915–87)
Studied the problem of rejection in tissue transplants, and discovered that the immune system learns to tolerate cells encountered at a young age.

Mendel, Gregor
Austrian monk & geneticist (1822–84)
Carried out experiments with plants to show how characteristics are inherited; his research helped to found the science of genetics.

Meselson, Matthew
American biochemist (born 1930)
Investigated how DNA copies itself, or replicates. Meselson showed that each of the two strands in DNA forms a new partner strand, instead of each complete molecule forming a copy of itself.

Meyerhof, Otto
German-American biochemist (1884–1951)
Discovered how lactic acid is formed in muscles.

Monod, Jacques
French biochemist (1910–76)
Discovered a mechanism that controls the way genes are turned on and off.

Montagu, Lady Mary
British writer (1689–1762)
Introduced inoculation against smallpox into England after seeing it successfully carried out in Turkey, where it had been in use for several centuries. Mary Montagu had her children successfully inoculated about 50 years before Edward Jenner (see page 101) demonstrated the value of the treatment. Smallpox was eventually eradicated as a result of inoculation.

Morgagni, Giovanni
Italian pathologist (1682–1771)
Investigated the causes of diseases by carrying out autopsies, and in 1761, published an important textbook on the subject.

Morgan, Thomas
American geneticist (1866–1945)
Developed the theory that chromosomes carry genetic information.

Müller, Johannes
German physiologist (1801–58)
Pioneer physiologist; investigated the senses, circulation, and nervous system.

◄ *Continued from previous page*

...ré, Ambroise
...ench surgeon (1510–90)
...oneered the use of dressings in
...lping wounds to heal, and was
...e first surgeon to tie severed
...od vessels during operations.

...steur, Louis
...ench microbiologist (1822–95)
...oneer of the science of
...icrobiology; put forward the
...rm theory of disease, which
...plains how infectious diseases
...e caused by microorganisms.

...uling, Linus
...nerican biochemist (born 1901)
...vestigated protein structure,
...d helped to shape ideas about
...e structure of DNA.

...vlov, Ivan
...ssian physiologist (1849–1936)
...howed that reflexes can be
...odified or created by learning.

...erutz, Max
...strian biochemist (born 1914)
...scovered the structure of
...moglobin.

...nel, Phillipe
...ench psychiatrist (1745–1826)
...oneered new methods in the
...eatment of mental illness. Until
...s time, most patients were
...ained up in prisonlike cells.

...ott, Percival
...ritish surgeon (1714–88)
...scovered that chimney sweeps
...n develop cancer by being
...xposed to soot. Pott's work was
...e first scientific study showing
...at some diseases are triggered by
...ctors in the environment.

...urkinje, Johannes
...zech cell biologist (1787–1869)
...ade detailed observations of
...lls, and discovered highly
...ranching neurons in the brain
...urkinje cells).

...amón y Cajal, Santiago
...panish physiologist (1852–1934)
...oneer of the study of nerves, and
...f nerve staining techniques. He
...as the first person to suggest
...at we learn by forming new
...onnections between neurons.

...éaumur, René Antoine
...ench naturalist (1683–1757)
...roved that chemicals were involved
...n digestion.

Rhazes
Arab doctor (c.865–932)
Prolific writer of medical texts, based
in Baghdad. Several of his works
were translated into Latin, and
circulated throughout Europe in
medieval times. Wrote the first
accurate description of smallpox.

Roentgen, Wilhelm
German physicist (1845–1923)
Discovered X-rays, and showed their
effects when allowed to strike a
photographic film.

Ross, Ronald
British medical scientist (1857–1932)
Unraveled the life cycle of the
parasite that causes malaria.

Roux, Pierre
French bacteriologist (1853–1933)
Discovered that bacteria release
powerful poisons (toxins) that cause
the symptoms of some diseases.

Salk, Jonas
American microbiologist (born 1914)
In 1954, developed the first safe and
effective vaccine against polio.

Santorio Santorio
(See page 103)

Sanger, Frederick
British biochemist (born 1918)
Became the first person to work
out the amino acid sequence in a
protein – in this case, insulin.

Schwann, Theodor
German physiologist (1810–82)
Discovered Schwann cells, which
surround neurons. Also helped
devise the theory that states that all
living things are made of cells.

Semmelweiss, Ignaz
Hungarian doctor (1818–65)
Discovered that personal hygiene,
particularly washing hands, had a
dramatic effect in reducing puerperal
fever, a common disease of childbirth.

Smith, Theobald
American pathologist (1859–1934)
Discovered that a cattle disease
could be spread by ticks; this was the
first scientific confirmation that
infectious diseases could be spread by
animals that bite.

Snow, John
British doctor (1813–58)
Discovered the role of contaminated
water in the spread of cholera.

Stanley, Wendell
American biochemist (1904–71)
Showed that purified viruses could
be crystallized.

Starling, Ernest
(See page 81)

Sturtevant, Alfred
American geneticist (1891–1970)
Developed techniques for mapping
chromosomes, by identifying genes
that are usually inherited together.

Swammerdam, Jan
Dutch microscopist (1637–80)
In 1658, became the first person to
observe and describe red blood cells.

Szent-Györgi, Albert von
(See page 49)

Takamine, Jokichi
(See page 81)

Vesalius, Andreas
Belgian anatomist (1514–64)
Published De Humani Corporis
Fabrica, or "On the Fabric of the
Human Body," the first accurate
account of the structure of the body.

Virchow, Rudolph
German cell biologist (1821–1902)
Helped to establish cell theory,
which states that all living things are
made of cells, and that cells are
always produced by other cells. Also
helped to lay the foundations of the
science of pathology.

Watson, James
American biochemist (born 1928)
With Francis Crick, correctly
deduced that DNA has a double
helix structure. This breakthrough,
made in 1953, helped to show how
genes are passed on from one
generation to another.

Wöhler, Friedrich
German chemist (1800–82)
Showed that urea, an organic
compound, could be made from
inorganic materials. This disproved
the widely held belief that the
chemicals in living and nonliving
things were separate and different.

Yersin, Alexandre
Swiss bacteriologist (1863–1943)
Discovered the plague bacterium
(at the same time as Kitasato); made
a vaccine to prevent the disease.

Young, Thomas
(See page 71)

Index

The index gives the page number of every main entry and subentry in this book. For a subentry, the main entry under which it appears is given in brackets ( ). Tables and table entries are shown by the italic word *table*.

Acknowledgments

Dorling Kindersley would like to thank:
Kate Eagar, Rachael Foster, Carlton Hibbert, Christopher Howson, and Robin Hunter for their design assistance. Claire Watts for her editorial help. Helen Annan for modeling. Photography by Andy Crawford, Peter Chadwick, Geoff Dann, Phillip Dowell, Philip Gatward, Steve Gorton, David Johnson, Dave King, David Murray, Dave Rudkin, Jules Seumes, Jane Stockman, Clive Streeter, and Matthew Ward.

David Burnie would like to thank the people whose work has contributed to the preparation of this book. Particular thanks go to Fiona Robertson, Gillian Cooling, Mark Regardsoe and the other members of the Dorling Kindersley team, for patience and enthusiasm shown throughout many months of hard work, and also to Richard Walker, for his invaluable scrutiny of the text and illustrations.

PICTURE CREDITS
l = left   r = right   t = top   b = bottom
c = center   a = above
Action Plus: / Chris Barry 56tl
Biophoto Associates: 74bl; 89cr; 96tr
Bruce Coleman: / Kim Taylor 93tl; 101cb;
Mary Evans Picture Library: 7bl; 49tr;
85tr; 93tr; 101tr; 103tr; 105tr
Hulton Deustch Collection: 71 tr
Image Select / Ann Ronan Collection: 11
Mansell Collection: 10; 77tr; 131cr
Microscopix: / Andrew Syred 100 br; 101tl
Oxford Medical Illustration: 80trb; / Carol

Barnett 100tr
Science Photo Library: 16 tr; 35br; 47br; /
Peter Aprahamian 23 br; / Biophoto
Associates 134br; 132b; / Dr Arnold Brody
113b; / Dr Jeremy Burgess 33tc; / Dr R Clark
& Mr R Goff 17tc; / CNRI 6cr; 28cr; 44bl;
82b; 83tl; 95bl; 95c; 125tr ; Prof. C Ferland
111cra; 111crb; Secchu Le Caque, Roussel-
Uglaf 9 cb; 59ct; / Custom Medical Stock
Photo, Robert Becker 12-13; John Smith
16c; / Martin Dohrn 68cbl; 68cbr; / Simon
Fraser 85c, Royal Victoria Infirmary,
Newcastle-upon-Tyne 126br; / John Greim
12tc; / Adam Hart-Davis 71cl; / Institut
Pasteur 29cr; 99cl; 99cr; / Manfred Kage
123t; 124br; Nancy Kedersha, Immunogen
31br; / Lungrafix 90tr; / Prof. R. Motta,
Dept. of Anatomy, University La Sapienza,
Rome 30tr; 75bl; 121cb; 130br; / NIBSC
16bl; 82tr; / Profs. P.M. Motta & S Correr
88b; / Alfred Pasteka 5tl; 26tl; / Dr Steve
Patterson 93cb; / Petit Format 7cl; 137b;
137cr; / D Philips 84bl; 136b; / K.R. Poiter
27cr; / J.C. Revy 37c; 80br; /Salisbury
District Hospital, Dept. of Clinical
Radiology 41tr; 47c; / Dr. K.F. R. Schiller
120br; / Jonathan Watts 93c.
Royal College of Physicians: 9c; 59tr;
81 tr; 113 ; 129tr
Sporting Pictures: 77cl
Tony Stone Images: 95br; 119br; / Lori
Adamski Peek 103 tl; / Doug Armand 81br; /
Bruce Ayres 93br / David Hiser 76bl; /
Laurence Monneret 138cl; / RNHRD NHS
Trust 42c; 87tr; 99cbl
 M.I. Walker: 48br; 48cl

Dr. R.M. Youngson: 69tl
ZEFA: 11bl; 11cl; 12bl; 33cl; 67tr; 89bl;
139bl; 140tl; 141cl; / Heilman 9cl; 59cr; /
Stockmarket 64tl

Every effort has been made to trace the
copyright holders. Dorling Kindersley
apologizes for any unintentional omissions
and would be pleased, in such cases, to add
an acknowledgment in future editions.

ARTWORK CREDITS
Peter Dennis: 142
DK Multimedia: 11br; 18bl; 51rc; 61c;
74tc; 75br; 79; 91; 96l; 97b; 110lc; 111lb;
126lc; 140c
William Donohue: 44/45b
Fernando Farah: 20/21; 22tl; 22tr; 23bl
Nick Hall: 48l
Janos Marffy: 25; 67br; 74rc; 84br; 87c;
102; 103c; 104; 105; 116b; 117bc; 119t;
120tr; 127; 132/133; 134bl; 135 tl;
138 rc; 139c
Marks Creative: 52tr; 53bl; 54t; 56r; 57l
Colin Salmon: 44rc; 45t; 62; 63; 90
Raymond Turvey: 69br
Lydia Umney: 18tl; 18bc; 18cr; 19lc; 19br
78; 114; 115; 122tl; 122bc; 123; 131
Peter Visscher: 34; 58; 60; 61br; 70t; 71bc
72tl; 76tc; 76br; 84tl; 88; 94br; 120bc; 121
137tl; 139tl
John Woodcock: 14br; 15lc; 15tr; 24b;
30rc; 31lc; 31c; 33r; 35c; 37cr; 39t; 41c;
55bl; 65c; 66tr; 73tc; 75tl; 82t; 83br; 86br;
87c; 96rc; 98; 100; 107; 118; 124; 125;
128bl; 136lc; 134tr; 141